THE 2011 NATIONAL ELECTRICAL CODE BOOK OF IN-DEPTH CALCULATIONS

VOLUME 1

Exclusively covering Article 220

Alvin J. Walker

THE 2011 NATIONAL ELECTRICAL CODE BOOK
OF IN-DEPTH CALCULATIONS

VOLUME 1

© 2015 Alvin J. Walker

ISBN 13: 978-0-9831358-2-1

LCCN 2015907070
First Edition
1 2 3 4 5 6 7 8 9 10

Walker & Walker Electrical Consultants
For more information, please contact
Alvin Walker
318-393-6841
www.alvinwalker.com

TABLE OF CONTENTS

Number in brackets indicates the number of questions per NEC sections. Total questions: 131

ARTICLE 220
Branch-Circuit, Feeder, and Service Calculations

PART I - General

PART II - Branch-Circuit Load Calculations

PART III - Feeder and Service Load Calculations

Standard Load Calculations

PART IV - Optional Feeder and Service Load Calculations

PART V - Farm Load Calculations

ACKNOWLEDGEMENTS

Although, this journey has been long and quite cumbersome along the way I am encouraged by the most profound and enlightening words I've ever known and that being the *Word of God*. For these precious words have given me the strength and determination to know that,

The race is neither given to the swift, nor the strong but to those who endure to the end.

In recognition, I would like to first give full honor and the biggest "thank you" to the Almighty God I serve, the Lord and Savior of my life, Jesus Christ and the empowerment of the Holy Spirit who has kept me focused and steadfast. And to my family and well-wishers who have supported me with much love and confidence.

But seek ye first the Kingdom of God, and His Righteousness;
and all these things shall be added unto you.
Matthew 6:33

PREFACE

The idea of writing a book such as this began over 11 years ago. It started off small and grew over the years. Once I began putting it together, I realized the enormity of the contents necessary to ensure the user of a well-rounded, in-depth reference. Never did I imagine the amount of time, effort and obstacles that I would encounter to produce a book that would offer such a sound means of understanding and performing electrical calculations. My intent, as in all previous books was to produce a clear and thorough source of information that would totally engage the user in all requirements of the National Electrical Code (NEC), while meticulously supporting the learning of the projected subject. Because of this, I pray that the intent, purpose and finished product are well- received and will be of great use.

As a whole, this book is broken down into four volumes.
- **Volume 1** consists solely of Article 220 which covers in great detail every aspect of calculating branch-circuit, feeder and service loads.
- **Volume 2** covers those articles of Chapters 1 through 3 that require either little or more profound electrical calculations. In **Volume 2**, the user will be intrigued by the clarity and practicality of how the more cumbersome articles are disclosed.
- The articles of Chapters 4 through 7 and the tables of Chapter 9 comprise **Volume 3**. **Volume 3** is an extension of **Volume 2** in that it also offers typical calculations that are common to everyday use.
- To assist the user in performing all possible calculations of the NEC, **Volume 4** contains a variation of eight detailed, step-by-step worksheets that will, for certain, enhance the user's confidence and capability for deriving reliable calculations, decisive conclusions and NEC compliance.

Undoubtedly, this volume of books provides an assuring pathway and a great reference that the user will find both rewarding and of immense value regardless of one's background or experience—whether an electrician, professional engineer, contractor, designer, salesperson or installer.

In considering the overall layout of **Volumes 1 - 3**, the following color scheme was used to provide a consistent path for quick reference of selected material where:
- red identifies those articles of the NEC involving electrical calculations.
- the color green is used to identify each specific section of an article that requires some type of an electrical calculation.
- followed by the color blue which is used to identify questions that are relative to each given section.
- Answers and explanations in response to each question are identified in black.
- Where there is a cross-reference of related questions deep-purple is used.
- Finally, the color brown is used to identify NEC material, supplements, discussions and pertinent information.
- Also of important note, where sections contain multiple questions, the relative question numbers are enclosed in parenthesis ( ), followed by the total number of questions in brackets [ ]. Volume 2 also features large-print for easier visibility and reading.

Combined, the first three volumes include more than 625 practical, yet everyday questions, along with detailed answers supported by an assortment of relative illustrations.

With this volume of books serving as the primary source along with other available books listed on our website, the opportunity to fill any void of understanding and learning the NEC can be quickly filled. For those individuals, groups, organizations, learning institution, etc. in search of material that produces immediate results and credible assistance--whether for NEC training/development, technical support or preparing for an electrician's examination at all levels--the above books are excellent sources.

INTRODUCTION TO VOLUME 1

Article 220 is the only Article of the National Electrical Code (NEC) which specifically uses the term "calculations". As titled, Article 220 is the exclusive source for referencing how to calculate branch-circuit; feeder and service loads which yield a conductor or set of conductors that are adequately sized to safely allow the expected amount of current flow. In summary, Article 220 is the total essences of Volume 1.

In Volume 1, the user will find that each featured section of Article 220 is uniquely arranged and clearly illustrated to render an in-depth comprehension of the subject matter to the user.

Filled with an assortment of thought-provoking questions and answers, Volume 1 will undoubtedly leave the user with both a complete understanding and greater appreciation of the NEC, in its effort to produce safe, reliable electrical systems.

ARTICLE 220 - Branch-Circuit, Feeder, and Service Calculations

As stated in NEC 220.1, the scope of Article 220 provides requirements for calculating branch-circuit, feeder, and service loads. Consisting of five parts, the intended purpose of Article 220 is to assist the user in computing electrical loads for residential, commercial, and industrial applications. Part I provides general requirements for calculation methods while Part II provides calculation methods for branch-circuit loads. For methods involving feeder and service loads, Part III reference sections that are essential for developing standard calculations for all occupancies while Part IV covers procedures pertaining to optional calculations for specific occupancies; dwellings, schools, and new restaurants. In conclusion, Part V deals exclusively with farm load calculations.

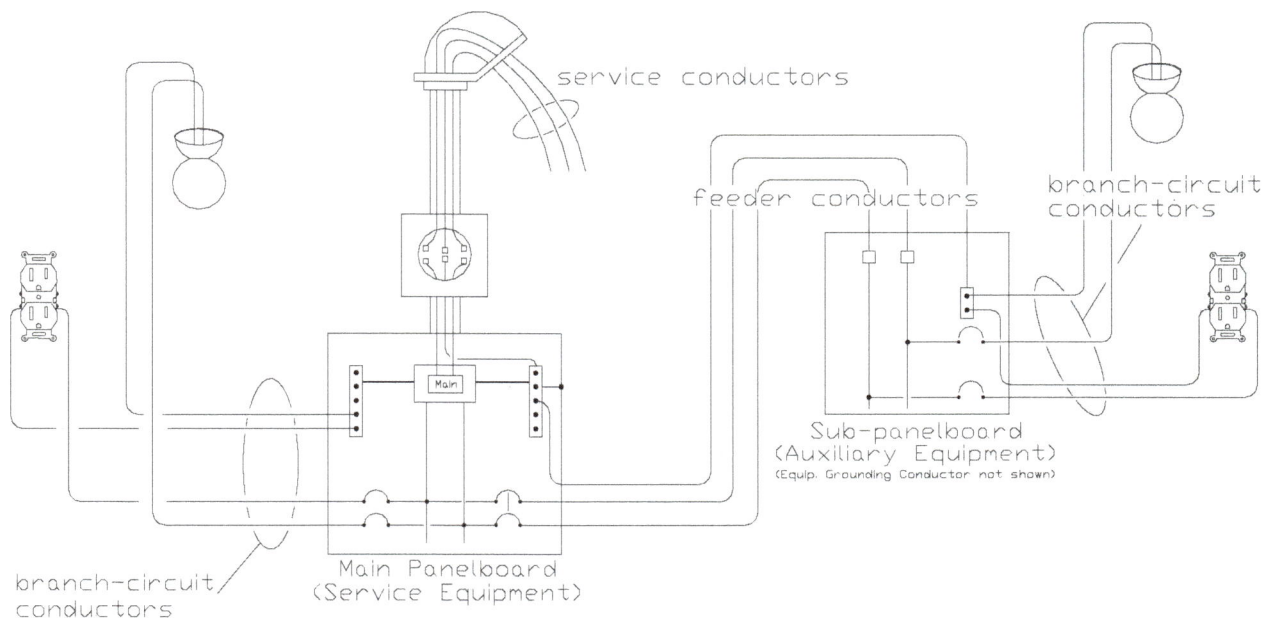

Figure 220.1 - Branch-circuit, Feeder and Service Conductors

PART I - General

220.5(B) - Fraction of an Ampere

1. Explain the permissive rule of NEC 220.5(B).

Although the rule is quite clear and self-explanatory the following example is used just in case.

A CALCULATION

$$87.4A \times 1.25 + 68A = 177.25A$$

APPLICATION

The permissive rule allows the fraction of the calculation to be dropped when resulting to a fraction that is less than 0.5 where in this calculation is .25A. The term fraction corresponds to a decimal. As a result, the rule allows the calculation to be expressed as 177A. However, on the other hand if the calculation resulted in a fraction that was greater than 0.5 that is, 177.52 then the fraction could be rounded up to a whole number which would leave the calculation at 178A.

In the questions and answers to follow - this permissive rule may not always be used.

PART II - Branch-Circuit Load Calculations

220.12 and **Table** - General Lighting Loads by Occupancy (2. - 7.) [6]

2. How many lighting, receptacle and smoke detector outlets are permitted on a 20A branch circuit supplying residential bedrooms?

A 120 volts, 20A branch circuit can supply a 2400VA (120V x 20A) load. Because the minimum load permitted in Table 220.12 is 3VA per square foot for dwelling units, one 20A branch circuit is allowed to supply an 800 square feet area (2400VA / 3VA/SF). For example, if a single dwelling has four bedrooms and the bedroom's combined square footage is either 800 square feet or less, a 20A branch circuit is permitted by code to supply all outlets mentioned although not recommended.

3. Determine the general lighting load that contributes to sizing service conductors supplying a 5800 square foot office building.

The unit load per square foot for an office building according to Table 220.12 is 3½VA per square foot. This value is used to determine the minimum general lighting load for the office building.

$$5800SF \times 3.5VA/SF = 20,300VA$$

Since this load is expected to operate continuously, NEC 230.42(A) requires the calculated value to be increased by 125 percent in determining the ampacity of service conductors. Therefore,

$$20,300VA \times 1.25 = 25,375VA$$

4. The interior lighting of an 115,000SF super store will consist of 1137 - 8' fluorescent light fixtures with 277V ballast rated for 1.15A each. Determine the store's general lighting load.

The general lighting load based on the unit load listed for a store provided in Table 220.12 is,

$$115,000SF \times 3VA/SF = 345,000VA$$

The actual (interior) lighting load based on the number of light fixtures is,

$$1137 \times 277V \times 1.15A = 362,191.35VA$$

Because the actual lighting load exceeds the computed general lighting load based on Table 220.12 it must be used. As you can see the minimum general lighting load for the store is still met and in compliance with the unit load per Table 220.12.

Since the store's service conductors must be sized to carry this load, NEC 230.42(A) requires the calculated value to be increased by 125 percent since the lighting load is expected to operate continuously for three hours or more. As a result,

$$362,191.35VA \times 1.25 = 452,739.2VA$$

The general lighting load for the super store is 452,739.2VA.

5. A motel has 2000 square feet of halls and stairs. How much lighting load must be added to consider these areas?

The unit load per square foot for halls and stairs according to Table 220.12 is ½VA per square foot. This value is used to determine the minimum general lighting load for halls and stairways.

$$2000SF \times \tfrac{1}{2}(.5)VA/SF = 1000VA$$

If the lighting loads in the halls and stairways of the motel are expected to operate continuously, the above load must be increased by 125 percent.

6. The number of receptacles in a 20,000 sq. ft. bank is unknown. Determine the receptacle demand load for this occupancy.

Table 220.12, Footnote "b" reference the use of NEC 220.14(K) as it applies to a bank. NEC 220.14(K) requires the receptacle loads for banks and office buildings to be calculated based on the larger of NEC 220.14(K)(1) [which reference the use of NEC 220.14(I)] or NEC 220.14(K)(2) [which reference the use of 1 volt-ampere/ft²]. Because the number of receptacles are unknown NEC 220.14(K)(1)[220.14(I)] cannot be applied. As a result, only NEC 220.14(K)(2) can be applied therefore, the receptacle demand load for the bank is,

$$20,000SF \times 1VA/SF = 20,000VA$$

Although the provisions of NEC 220.44 allows the use of demand factors for Non-Dwelling Receptacle Loads, they are not allowed under these conditions, only those receptacle loads calculated per NEC 220.14(H) and 220.14(I) are.

Refer to question No. 14. for office building receptacle loads.

7. How many 2-wire, 20 amperes, 120V receptacle circuits are required for question No. 6.?

Considering the volt-amperes demand of a 120 volts - 20 amperes circuit which is 2400VA (120 volts x 20 amperes) and the 20,000VA connected load, the number of 20 amperes circuits can be determined,

$$\frac{20{,}000VA}{120V \text{ x } 20A} = 8.3$$

Rounding the calculated value 8.3 up to the nearest whole number requires a minimum of 9-20 amperes circuits to supply the bank's receptacle loads.

Per NEC 220.14(I) at 180VA per receptacle, only 13 receptacles are allowed on 20A circuits (20A x 120V = 2400VA/180VA = 13.33-rounded down). Based on 180VA per receptacle, the bank could hypothetically contain a minimum of 111 receptacles (20,000VA/180VA).

220.14(D) - Luminaries

8. The label of a 6" recessed can incandescent light fixture rated for 120V indicates that the fixture is rated to supply bulbs up to 100 watts. If 18 fixtures are used, determine the branch-circuit demand load of the fixtures when used continuously or noncontinuously.

Although bulbs with a smaller wattage rating could be used in the fixtures, the branch-circuit demand load of the fixtures must be based on the fixture's maximum volt-amperes rating where in this case is 100 watts according to NEC 220.14(D) for noninductive light fixtures. As a result,

For *noncontinuous* operation the branch-circuit demand load is,

$$100W(VA) \text{ x } 18 = 1800VA$$

For *continuous* operation the branch-circuit demand load is,

$$100W(VA) \text{ x } 18 \text{ x } 1.25 = 2250VA$$

where the 125 percent (1.25) increase is required when sizing branch-circuit conductors and overcurrent devices per NEC 210.19(A)(1) and 210.20(A) or when sizing feeder and service conductors per NEC 215.2(A)(1) and 230.42(A)(1).

220.14(E) - Heavy-Duty Lampholders

9. Determine the branch-circuit load for 6 admedium type heavy-duty lampholders that's rated for 700 watts.

A heavy-duty lampholder is defined as one having a rating of not less than 660 watts if of the admedium type, and not less than 750 watts if of any other type. See NEC 210.21(A).

NEC 220.14(E) requires outlets for heavy-duty lampholders to be computed at a minimum of 600 volt-amperes. Because the lampholders described in the question have a rating that exceeds the minimum 600 volt-amperes rating the branch-circuit load is as calculated,

$$700 \text{ watts x } 6 = 4200 \text{ watts (volt-amperes)}$$

220.14(F) - Signs and Outline Lighting

10. An exterior commercial sign is rated for 1000W at 120V. What size branch circuit is required to supply the sign?

Although the amount of current drawn by the sign is 8.3A (1000W /120V) and when multiplied by 125 percent the value is increased to 10.38A [because of the continuous load requirements of NEC 210.20(A)], it appears that a circuit rated for 15 amperes can be used. However, NEC 220.14(F) requires sign and outline lighting outlets to be computed at a minimum of 1200 volt-amps and NEC 600.5(A) requires sign or outline lighting outlets to be supplied by a branch circuit that's rated for at least 20 amperes. Even when calculated at 1200VA and increased by 125 percent (1200VA/120V x 1.25 = 12.5A) the end results are less than 20 amperes. Nevertheless, the fact remains and that is, NEC 600.5(A) must be applied. Based on the use of a 20 amperes overcurrent device the branch-circuit conductors must be at least a 12 AWG copper for this application according to NEC 240.4(D)(5).

220.14(G) - Show Windows

11. How many 20 amperes branch circuits are needed to supply 42 feet of show window lighting in a men's clothing store? Consider both 120V and 277V circuits.

NEC 220.14(G) requires show window lighting to be computed at 200 volt-amperes per foot when determining the branch-circuit load. Therefore,

$$200\text{VA/ft x } 42\text{ft} = 8400\text{VA}$$

Because show window lighting is expected to operate continuously, the computed load of the branch circuit must be increased by 125 percent according to NEC 210.20(A) which now becomes,

$$8400\text{VA x } 1.25 = 10,500\text{VA}$$

Using this computed value the number of 20 amperes branch circuits can now be determined,

At 120V

$$\frac{10,500\text{VA}}{120\text{V x } 20\text{A}} = 4.375$$

When rounded up, 5-20 amperes branch circuits are required to supply the show window lighting at **120V**.

At 277V

$$\frac{10,500\text{VA}}{277\text{V} \times 20\text{A}} = 1.89$$

When rounded up, two 20 amperes branch circuits are required to supply the show window lighting at **277V**.

220.14(H) - Fixed Multioutlet Assemblies

12. Extended runs of plugmolds are installed in the laboratory of a testing facility to supply various types of appliances. For occasional testing, 58 feet of plugmolds are installed for such appliances. Where appliances are required to be tested simultaneously, 32 feet of plugmolds are installed. Determine the plugmolds' branch-circuit load.

Plugmolds are referred to as multioutlet assemblies in The National Electrical Code and can be referenced in Article 380. However, to calculate the branch-circuit load of such equipment NEC 220.14(H) is referenced. Commonly speaking, this reference requires the branch-circuit load of the plugmolds to be calculated at 180VA for every 5 feet (5') of the plugmold's assembly where the plugmolds are not used simultaneously (at the same time) and at 180VA for every foot (1') of the plugmold's assembly where the plugmolds are used simultaneously (simultaneous usage does not mean continuous duty). For non-simultaneous use of plugmolds where every 5 feet of use equals 180VA the load is,

$$180\text{VA}/5' \times 58' = 2088\text{VA}$$

for simultaneous use of plugmolds where every foot of use equals 180VA the load is,

$$180\text{VA} \times 32' = 5760 \text{ VA}$$

The plugmold's branch-circuit load is 7848VA (2088VA + 5760VA). Although NEC 220.44 permits the load of multioutlet assemblies (receptacles) to be reduced this only applies when the receptacle load exceeds 10kVA (10,000VA).

220.14(I) - Receptacle Outlets

13. How many duplex receptacles can be placed on a 15 or 20 amperes branch circuit in a nondwelling?

Unlike dwellings, where the number of receptacles (or lighting outlets) placed on a branch circuit has no limit per National Electrical Code (although not recommended), NEC 220.14(I) limits the number of receptacles placed on branch circuits in nondwellings, meaning commercial and industrial occupancies, to 180VA per yoke (strap) for each single or multiple numbers of receptacles on a yoke.

Because 15 and 20 amps (amperes) branch circuits feeding receptacle loads are supplied by a 120 volts source, the volt-amperes rating of a 15 amps branch circuit is 1800 volt-amperes (VA) (120 volts x 15 amps) and for a 20 amps circuit the volt-amperes rating is 2400 volt-amperes (VA) (120 volts x 20 amps).

Using both required and calculated information, the number of duplex receptacles that can be placed on a

15 amps branch circuit is **10** (1800VA / 180VA)

and on a

20 amps branch circuit is **13** [rounded down] (2400VA / 180VA).

220.14(K) - Banks and Office Buildings

14. A 4835SF office building will require 228 receptacles to be installed. Determine the receptacle load for the building.

NEC 220.14(K) requires the receptacle load for banks and office buildings to be determined based on the larger of either NEC 220.14(K)(1) where each receptacle is calculated at 180VA per NEC 220.14(I) or at 1 volt-ampere per square foot per NEC 220.14(K)(2). Considering the requirements of NEC 220.14(I) first where,

$$180VA \times 228 = 41,040VA$$

followed by the requirements of NEC 220.14(K)(2) where,

$$4835SF \times 1VA/SF = 4835VA$$

Clearly, the calculated results gathered per NEC 220.14(K)(1) is the larger of the two therefore, the receptacle load for the office building is 41,040VA.

However, unlike the situation found in question No. 6. where the provisions of NEC 220.44 were not applicable, the demand factors of Table 220.44 can be applied for this application.

Referring to Table 220.44, the receptacle demand load is determined as follows:

First 10kVA   = 10,000VA

  (41,040VA – 10,000VA = 31,040VA)

Remainder (31,040VA) at 50% (31,040VA x .50) = <u>15,520VA</u>
                                                    25,520VA

Based on the calculated results the receptacle demand load for the bank is 25,520VA. If the receptacle load was rated for continuous use the provisions of NEC 220.44 would not be applicable.

220.14(L) - Other Outlets

For those outlets loads not covered in NEC 220.14(A) - (K), NEC 220.14(L) requires such loads to be calculated at 180VA per outlet.

220.16 - Loads for Additions to Existing Installations

The provisions of NEC 220.16(A) and (B) re-emphasize the use of NEC 220.12 and NEC 220.14 for calculating branch-circuit loads where (1) structural additions (add-ons) occurs or (2) new or extended (remodeling) circuits are included in existing dwellings or non-dwelling units (commercial and industrial buildings).

220.18(A) - Motor-Operated and Combination Loads

See Article 430 (Volume 3) for circuits supplying only motor-operated loads. See Article 440 (Volume 3) for circuits supplying only air-conditioning equipment, refrigeration equipment, or both.

220.18(B) - Inductive and LED Lighting Loads

15. A 2' x 4' fluorescent light fixture contains 4-40 watt lamps. The lamps are supplied by two ballast rated for .35 amps each at 277V. If 54 fixtures are used, determine the total load in volt-amperes.

When circuits supplying lighting units are equipped with ballasts, NEC 220.18(B) requires the load of the fixture to be computed based on the total ampere ratings of the ballast(s) instead of the total watts of the lamps (160W). For 54 fixtures, the total load in volt-amperes is,

$$2 \times 277 \text{ volts} \times .35 \text{ amps} \times 54 \text{ fixtures} = 10,470.6\text{VA}$$

If these fixtures will be used in an area where continuous use is required the total load in volt-amperes is,

$$10,470.6\text{VA} \times 1.25 = 13,088.25\text{VA}$$

220.18(C) - Range Loads

See NEC 220.55 for questions pertaining to demand factors for "Range Loads".

PART III - Feeder and Service Load Calculations

UNDERSTANDING DEMAND FACTORS

As it pertains to an electrical system, the term *demand factor* is the ratio (or percentage) of that portion of the connected load that will operate at the same time *to* the total amount of the connected load. For example, a 50 percent demand factor implies that only 50 percent (or half) of the connected load will ever operate at the same time.

In Table 220.42 only four types of occupancies (dwelling units, hospitals, hotels/motels and warehouses) are listed where demand factors are permitted to be applied for general illumination (lighting load). For all other type occupancies general illumination is assumed continuous and requires being calculated at 125 percent for feeder [NEC 215.2(A)(1)] and service conductors [NEC 230.42(A)(1)]. Where general illumination is assumed to be noncontinuous it's calculated at 100 percent.

In summary, a demand factor only considers a portion of a load. A noncontinuous load implies that the load will be energized or in constant use for less than three hours. In comparison, a continuous load implies that the load will be energized or in constant use for three hours or more. When a demand factor is applied a load is never considered nor reconsidered as noncontinuous (at 100 percent) or continuous (at 125 percent).

As for other tables of the NEC where demand factors are permitted the same summarized provisions apply per application. For other NEC applications, refer to the following tables:

Table	NEC Application
220.44	Non-Dwelling Receptacle Loads
220.54	Electric Clothes Dryers—Dwelling Unit(s)
220.55	Electric Ranges and Cooking Appliances—Dwelling Unit(s)
220.56	Kitchen Equipment—Other Than Dwelling Unit(s)
220.84	Optional Calculations for Three or More Multifamily Dwelling Units
220.86	Optional Method for Schools
220.88	Optional Method for New Restaurants
220.102	Method for Calculating Farm Loads for Other Than Dwelling Unit
220.103	Method for Calculating Total Farm Load
530.19(A)	Stage Set Lighting
550.31	Mobile Homes Service and Feeders
551.73	Site Feeders and Service-Entrance Conductors for Park Sites
555.12	Service and/or Feeder Circuits for Boat Receptacles
610.14(E)	Cranes or Hoists
620.14	Elevators
626.11(B)	Electrified Truck Parking Spaces

Table 220.42 - Lighting Load Demand Factors (16. - 17.) [2]

16. Determine the service lighting load of a 158,000 square feet hospital where the lighting requirements in the emergency and operating rooms of the hospital are continuous and the total square footage of both areas is 7200 square feet.

The footnote to Table 220.42 states that the demand factors of the Table are not applicable where the entire lighting is likely to be used at one time (operating at the same time where continuous usage is expected). This being the case, the emergency and operating rooms must not be computed with the total square footage of the hospital to determine the lighting load demand per demand factors.

$$158,000SF - 7200SF = 150,800SF$$

Referring to Table 220.12 a unit load of 2VA per square foot is used for computing the general lighting load for a hospital.

$$150,800SF \times 2VA/SF = 301,600VA$$

The lighting load demand factors given in Table 220.42 for a hospital can now be applied.

First 50,000VA or less (At 40%)

$$50,000VA \times .40 = \textbf{20,000VA}$$

Remainder over 50,000VA (At 20%)   (301,600VA – 50,000VA = 251,600VA)

$$251,600VA \times .20 = \textbf{50,320VA}$$

Total based on use of demand factors

$$20,000VA + 50,320VA = 70,320VA$$

Continuous loads (Emergency and Operating Rooms) per NEC 230.42(A)(1)

$$7,200SF \times 2VA/SF \times 1.25 = 18,000VA$$

The service general lighting load for the hospital is 88,320VA (70,320VA + 18,000 VA).

17. Determine the service lighting load of a 122,000 square feet hotel where provisions for cooking are not provided. The following areas are included in the hotel and are expected to require lighting no less than 8 hours a day:

> 3650 square feet entertainment and festivity area
> 4000 square feet dining facility
> 1700 square feet of meeting rooms

1000 square feet health spa
450 square feet laundry facility

Just as in question No. 16., the footnote to Table 220.42 must be adhered to. For those areas of the hotel where lighting is expected to operate for 3 hours or more the lighting loads must be computed separately.

Total square feet of areas where continuous lighting is expected: 10,800 SF

$$122,000SF - 10,800SF = 111,200 \text{ SF}$$

Square footage used to compute lighting demand per Table 220.42: 111,200 SF.

Referring to Table 220.12 a unit load of 2VA per square foot is also used for computing the general lighting load for hotels.

$$111,200SF \times 2VA/SF = 222,400 \text{ VA}$$

The lighting load demand factors given in Table 220.42 for a hotel is,

---

First 20,000VA or less (At 50%)

$$20,000VA \times .50 = \textbf{10,000VA}$$

From 20,001VA to 100,000VA (80,000VA) (At 40%)

$$80,000VA \times .40 = \textbf{32,000VA}$$

Remainder over 100,000VA (At 30%) (222,400VA − 100,000VA = 122,400VA)

$$122,400 \text{ VA} \times .30 = \textbf{36,720VA}$$

Total based on use of demand factors

$$10,000VA + 32,000VA + 36,720VA = 78,720VA$$

Continuous loads (Other Areas) per NEC 230.42(A)(1)

$$10,800SF \times 2VA/SF \times 1.25 = 27,000 \text{ VA}$$

The service general lighting load for the hotel is 105,720VA (78,720VA + 27,000 VA).

220.43(A) - Show Windows

18. What portion of a jewelry store's show window lighting will contribute to the store's feeder demand load if the show window's total lighting load covers a 77 feet area?

Just as required in NEC 220.14(G) for show window branch circuits, NEC 220.43(A) requires show window lighting to be computed at 200 volt-amperes per foot to determine the computed load feeder circuits. Therefore,

$$200VA/ft \times 77ft = 15,400VA$$

Because show window lighting is expected to operate continuously, the feeder demand load must be increased by 125 percent according to NEC 215.2(A)(1) which results in the following,

$$15,400VA \times 1.25 = 19,250VA$$

The show window lighting will contribute 19,250VA towards the store's feeder demand load.

220.43(B) - Track Lighting

19. Two rows of track lighting measuring 44 feet each in length are to be installed in a shoe store to display a large selection of men's and women's shoes. Determine the service demand of the track lighting.

NEC 220.43(B) requires track lighting to be computed at 150 volt-amperes for every 2 feet (2') of track lighting in determining the service demand of the lighting. Therefore,

$$150VA/2ft \times 44ft \times 2(rows) = 6600VA$$

Because the track lighting is expected to operate continuously, the service demand load must be increased by 125 percent according to NEC 230.42(A)(1) which results in the following,

$$6600VA \times 1.25 = 8250VA$$

The service demand for the track lighting is 8250VA. This calculated load only identifies the track lighting load that will contribute to the total service demand load once all other loads are considered.

220.44 and **Table** - Receptacle Loads – Other Than Dwelling Units (20. - 22.) [3]

Receptacles loads calculated in accordance with NEC 220.14(H) and (I) are permitted to be made subject to the demand factors given in Table 220.44.

20. Determine the receptacle load for a small convenient store which has 34 duplex receptacles.

Per NEC 220.14(I), receptacle outlets in nondwelling units are required to be computed at 180 VA per outlet where,

$$180VA \times 34 = 6120VA$$

The receptacle load for the store is 6120VA per Table 220.44.

REMINDER - The demand factors in Table 220.44 will not apply until 56 or more receptacles (at 180VA) are utilized or the total receptacle load exceeds 10,000VA.

21. An office has a total of 373 receptacles. Determine the receptacle load for the office.

$$180VA \times 373 = 67,140VA \ (67.14kVA)$$

Because the computed value exceeds 10kVA the demand factors in Table 220.44 can be applied.

First 10kVA (10,000VA) or less (At 100%)
10,000VA
Remainder over 10,000VA (At 50%)   (67,140VA – 10,000VA = 57,140VA)
57,140VA x .50 = **28,570VA**

The receptacle load for the office is 38,570VA (10,000VA + 28,570VA).

22. A 10,000 square feet storage warehouse has 125 duplex receptacles and 137 feet of multioutlet assemblies where most of the outlets are used at the same time. Determine the service demand of the warehouse applying the provisions of NEC 220.44.

To determine the service demand for this occupancy instead of calculating the general lighting load and receptacle loads separately, NEC 220.44 permits the loads to be calculated together and made subject to the demand factors given in Table 220.42. As an alternative, NEC 220.44 also allows the receptacle loads to be calculated per Table 220.44 where in this case the general lighting load and receptacle loads are calculated per Tables 220.42 and 220.44 respectively.

Now let's calculate the service demand load for the warehouse considering both general lighting and receptacle loads together per Table 220.42.

To determine the service demand for the warehouse, the general lighting load must first be computed. Per Table 220.12 the unit load for a warehouse is ¼VA per square foot.

General Lighting Load [Table 220.12]

$$10,000SF \times ¼VA/SF = 2,500VA$$

Receptacle Loads [NEC 220.14(I)]

$$180VA \times 125 = 22,500VA$$

Multioutlet Assemblies [NEC 220.14(H)(2) – [180VA/ft.] simultaneous use]

$$180VA \times 137' = 24,660VA$$

The total load to be computed per demand factors is 49,660VA (2,500VA + 22,500VA + 24,660VA).

First 12,500VA or less (At 100%)

12,500VA

Remainder over 12,500VA (At 50%)   (49,660VA – 12,500VA = 37,160VA)

37,160VA x .50 = **18,580VA**

The service demand for the warehouse using the provisions of NEC 220.42 is 31,080VA (12,500VA + 18,580VA).

Now let's calculate the service demand load for the warehouse considering both general lighting and receptacle loads separately per Table 220.44.

The total receptacle and multioutlet assemblies loads total 47,160VA (22,500VA + 24,660VA). Applying the total receptacle load to the demand factors of Table 220.44,

---

First 10kVA (10,000VA) or less (At 100%)

10,000VA

Remainder over 10,000VA (At 50%)   (47,160VA – 10,000VA = 37,160VA)

37,160VA x .50 = **18,580VA**

Based on the calculated results per Table 220.44 the receptacles demand load is 28,580VA (10,000VA + 18,580VA). Considering the general lighting load at 2,500VA and the receptacle demand load per Table 220.44 the service demand for the warehouse is 31,080VA (28,580VA + 2,500VA).

As you can see the results based on the provisions of Tables 220.42 and 220.44 are the same for both applications, 31,080VA.

220.51 - Fixed Electric Space Heating

23. A 20kW electric heating unit along with a 15kW unit is used to provide heating for a 4000 square feet building. Determine the service demand for the heating units.

Where fixed electric space heating is installed, NEC 220.51 requires such loads to be computed at 100 percent of the total connected load. For this installation the service demand based on the heating units is 35kW (20kW + 15kW).

220.52(A) - Small-Appliance (Portable) Circuit Load (24. - 25.) [2]

24. Five small-appliance circuits will be used in a 6500 square feet dwelling. Determine the connected load of the small-appliance circuit loads before and the demand load after the demand factors of Table 220.42 are applied.

To compute the small-appliance circuit load for a dwelling, NEC 220.52(A) requires each small-appliance circuit to be computed at 1500 volt-amperes (VA). NEC 220.52(A) also permits small-appliance circuit loads to be included with the general lighting load and subjected to the demand factors provided in Table 220.42.

Connected load before demand factors

$$1500VA \times 5 = 7500VA$$

Demand load after demand factors

Per Table 220.42, for this dwelling is 4575VA (3000VA + 1575VA).

First 3000VA or less (At 100%)
3000VA
From 3001VA to 120,000VA (7500VA – 3000VA = 4500VA) (At 35%)
4500VA x .35 = **1575VA**

25. Determine the small-appliance service demand load for a 75 unit apartment complex.

In accordance with NEC 210.11(C)(1), a minimum of two 20-amperes small-appliance branch circuits are required for dwelling units. Again, NEC 220.52(A) requires small-appliances loads to be calculated at 1500 volt-amperes per branch circuit. Considering both requirements the small-appliance service demand for a 75 unit apartment complex is calculated as follows:

$$75 \times 1500VA \times 2 = 225,000VA \text{ (connected load)}$$

As permitted in NEC 220.52(A) the demand factors of Table 220.42 are applied for dwelling units.

First 3000VA or less (At 100%)
3000VA
From 3001VA to 120,000VA (120,00VA – 3000VA = 117,000VA) (At 35%)
117,000VA x .35 = **40,950VA**

Remainder over 120,000VA (225,00VA – 120,000VA = 105,000VA) (At 25%)

105,000VA x .25 = **26,250VA**

The small-appliance service demand for the 75 unit apartment complex is 70,200VA (3000VA + 40,950VA + 26,250VA).

220.52(B) - Laundry Circuit Load (26. - 27.) [2]

26. What is the demand load for a laundry circuit load in a single-family dwelling?

NEC 220.52(B) requires the demand load for a laundry circuit in a dwelling to be computed at no less than 1500 volt-amperes (VA). Therefore, the laundry circuit demand load for this dwelling in 1500VA.

Although the demand factors of Table 220.42 are permitted per NEC 220.52(B), in order to apply this provision the laundry demand load would have to exceed 3000VA.

27. Determine the laundry service demand load for a 75 unit apartment complex.

In accordance with NEC 210.11(C)(2) a minimum of one 20-amperes laundry branch circuit is required for dwelling units. Again, NEC 220.52(B) require laundry loads to be calculated at 1500 volt-amperes per branch circuit. Considering both requirements the laundry service demand for a 75 unit apartment complex is calculated as follows:

75 x 1500VA = 112,500VA (connected load)

As permitted in NEC 220.52(B) the demand factors of Table 220.42 are applied for the dwelling units.

First 3000VA or less (At 100%)

3,000VA

From 3001VA to 120,000VA (112,500VA – 3000VA = 109,500VA) (At 35%)

109,500VA x .35 = **38,325VA**

The laundry service demand for the 75 unit apartment complex is 41,325VA (3000VA + 38,325VA).

220.53 - Appliance (Fastened in Place) Load – Dwelling Unit(s) (28. - 30.) [3]

The small-appliance circuit load per NEC 220.52(A) pertains to portable household appliances such as, toasters, coffee-makers, can-openers, blenders, etc. whereas, NEC 220.53 pertains to permanently fixed household appliances such as dishwashers, garbage disposals, trash compactors, etc.

28. The following appliances are installed in a one-family residence:

<table>
<tr><td colspan="2">240/120V</td><td colspan="2">120V</td></tr>
<tr><td>7.2kW Wall-mounted oven</td><td></td><td>1kW Dishwasher</td><td></td></tr>
<tr><td>7.4kW Cooktop</td><td></td><td>1750W Microwave oven</td><td></td></tr>
<tr><td>15kW Electric heat</td><td></td><td>1300W Heat-Vent-Light (3)</td><td></td></tr>
<tr><td>26kVA AC</td><td></td><td>960VA Disposal</td><td></td></tr>
<tr><td>1320VA Blower motor</td><td></td><td>720VA Trash compactor</td><td></td></tr>
<tr><td>4.5kW Dryer</td><td></td><td></td><td></td></tr>
<tr><td>6000VA Indoor spa</td><td></td><td></td><td></td></tr>
</table>

240V
5kW Water heater (2)
8kVA A/C unit

Determine the service demand for the appliances.

When four (4) or more appliances (other than electric ranges, clothes dryers, space-heating equipment, or air-conditioning equipment) are fastened in place (fixed) in a dwelling unit, NEC 220.53 permits a 75 percent demand factor to be applied towards the nameplate rating of each appliance. Excluding the equipment which the demand factor cannot be applied, the following appliance loads are permitted for derating:

(1) Dishwasher -	1000VA
(1) Microwave oven -	1750VA
(3) Heat-Vent-Light -	3900VA
(1) Disposal -	960VA
(1) Trash compactor -	720VA
(2) Water heater -	10,000VA
(1) Indoor spa -	6000VA
(10) Appliances	24,330VA

Although it may not appear as obvious as others, the indoor spa is considered an appliance based on the definition provided in Article 100. Also refer to NEC 680.2 for the definition of a spa and other similar equipment. The blower motor was not listed as an appliance because it is a component of the heating and air-conditioning equipment. Applying the demand factor,

$$24,330VA \times .75 = 18,247.5VA$$

The service demand for the appliances is 18,247.5VA. Unlike small-appliances the appliances referenced in NEC 220.53 are fixed appliances and usually require being placed on individual branch circuits. The small-appliances referenced in NEC 220.52(A) require no individual calculations because they are supplied by either of the two minimum small-appliance circuits or more. Small appliances are portable (not fastened in place) meaning they can be moved from one small-appliance receptacle or circuit to another.

29. Determine the neutral load for the appliances in question No. 28.

Of all permitted appliances in question No. 28., the water heaters are the only appliances not requiring neutral connections because they are solely line-to-line loads requiring only line-to-line connections. As a result, both appliance loads must be subtracted from the total appliance loads with neutral connections followed by the application of the 75 percent demand factor. With the two water heater loads subtracted there are still 8 appliances remaining which still exceeds the 4 appliance minimum to apply the 75 percent demand factor.

$$24,330VA - 10,000VA = 14,330VA \text{ (based on 8 appliances)}$$

$$14,330VA \times .75 = 10,747.5VA$$

The neutral load for the appliances is 10,747.5VA.

30. A 25 unit apartment complex provides whirlpool tubs in each unit. The tubs have an 1800VA nameplate rating. Determine the appliance demand load of the whirlpool tubs.

$$1800VA \times 25 \times .75 = 33,750VA$$

The appliance demand load for the whirlpools is 33,750VA.

220.54 and **Table** - Electric Clothes Dryers - Dwelling Unit(s) (31. - 34.) [4]

In accordance with NEC 220.54, kilovolt-amperes (kVA) [or volt-amperes (VA)] shall be considered equivalent to kilowatts (kW) [or watts (W)] for loads calculated in this section.

Refer to **(Worksheet E** - Volume 4) DEMAND LOAD CALCULATIONS FOR HOUSEHOLD ELECTRIC DRYERS for related questions.

31. Determine the demand load for a household electric clothes dryer with a nameplate rating of 6.3kW?

According to NEC 220.54, the load for household electric clothes dryers in dwelling units must be rated for 5000 watts (volt-amperes) or the dryer's nameplate rating whichever is larger. In this case the demand load for the dryer is 6300 watts, the larger of the 5000 watts minimum requirement.

32. What is the service demand load in amperes for 8 household electric dryers rated for 4.7kW at 208/120V? The dryers will be fed from a single-phase system.

Because the dryers are rated (nameplate) for 4.7kW (4700 watts) the load of each dryer must be calculated at 5000 watts to be code compliant. Therefore, ~~4700~~ 5000 watts x 8 = 40,000 watts.

The 40,000 watts (calculated value) represents the connected load of the 8 dryers. To determine the service demand load for the dryers, the demand factors of Table 220.54 are permitted. As listed in the table the demand factor for 8 dryers is 60 percent (.60).

$$40,000 \text{ watts x } .60 = 24,000 \text{ watts}$$

The 24,000 watts calculated represents the service demand load for the 8 dryers. Now the only thing left to do is to convert the load in watts (W) into amperes.

$$24,000W / 208V = 115.38 \text{ amps}$$

The service demand load for the 8 dryers at 208/120V-1ϕ is 115.38 amps.

33. Determine the service demand load for each set of multifamily dwelling units based on the ratings of the single-phase electric dryers each set of units are equipped with.

17 units - 4.25kW dryers   40 units - 5kW dryers   65 units - 5.6kW dryers

Because one unit of dryers are rated for 4.25kW (4250W), NEC 220.54 requires the calculated load for all household electric dryers to be no less than 5000W. Since the ratings of the other dryers are either equal to or exceeds 5000W they can be calculated as is,

~~4250~~ 5000 watts x 17 = 85,000 watts   5000 watts x 40 = 200,000 watts
5600 watts x 65 = 364,000 watts

The calculated results represent the connected load for each unit of dryers. To determine the service demand load for the dryers, the demand factors of Table 220.54 are permitted. As listed in the table, the demand factor for the dryers must be based on the given formulas therefore,

For **17 units** - Demand Factor = 47% – [1% x (17 – 11)]
= 47% – 1% x (6)
= 47% – 6%
= 41% or .41
85,000 watts x .41 = **34,850 watts**
The service demand load for the **17 dryers** is **34,850 watts.**

For **40 units** - Demand Factor = 35% – [.5% x (40 – 23)]
= 35% – .5% x (17)
= 35% – 8.5%
= 26.5% or .265
200,000 watts x .265 = **53,000 watts**
The service demand load for the **40 dryers** is **53,000 watts.**

For **65 units** - Demand Factor = 25 percent (.25)
364,000 watts x .25 = **91,000 watts**
The service demand load for the **65 dryers** is **91,000 watts.**

Single-Phase Dryers on Three-Phase, 4-Wire System

34. Determine the service demand load in amperes for the dryers in question No. 33. if the dryers were connected to a 3-phase, 4W, 208/120V service.

According to NEC 220.54, where two or more single-phase dryers are supplied by a 3-phase, 4-wire service, the total load of the single-phase dryers must be computed on the basis of twice the maximum number of dryers connected between any two phases. Observe how this requirement is interpreted and applied.

The 3-phase, 4-wire service referenced in NEC 220.54 refers to a 3-phase, 4-wire, 208/120V wye-connected source. In a 3-phase, 4-wire, wye-connected system there are two phase windings between each line-to-line connection which means this occurs three times. So now let's determine the maximum number of dryers connected between any two phases by dividing the number of dryers by the number of phases and rounding up to the nearest whole number if needed,

$$\frac{17 \text{ dryers}}{3} = (5.67) = \mathbf{6} \qquad \frac{40 \text{ dryers}}{3} = (13.3) = \mathbf{14} \qquad \frac{65 \text{ dryers}}{3} = (21.67) = 22$$

Since 6, 14 and 22 dryers are the maximum number that can be connected between any two phases; the remaining dryers will be evenly divided between the other sets of two phase windings. See Figure 220.54-34.

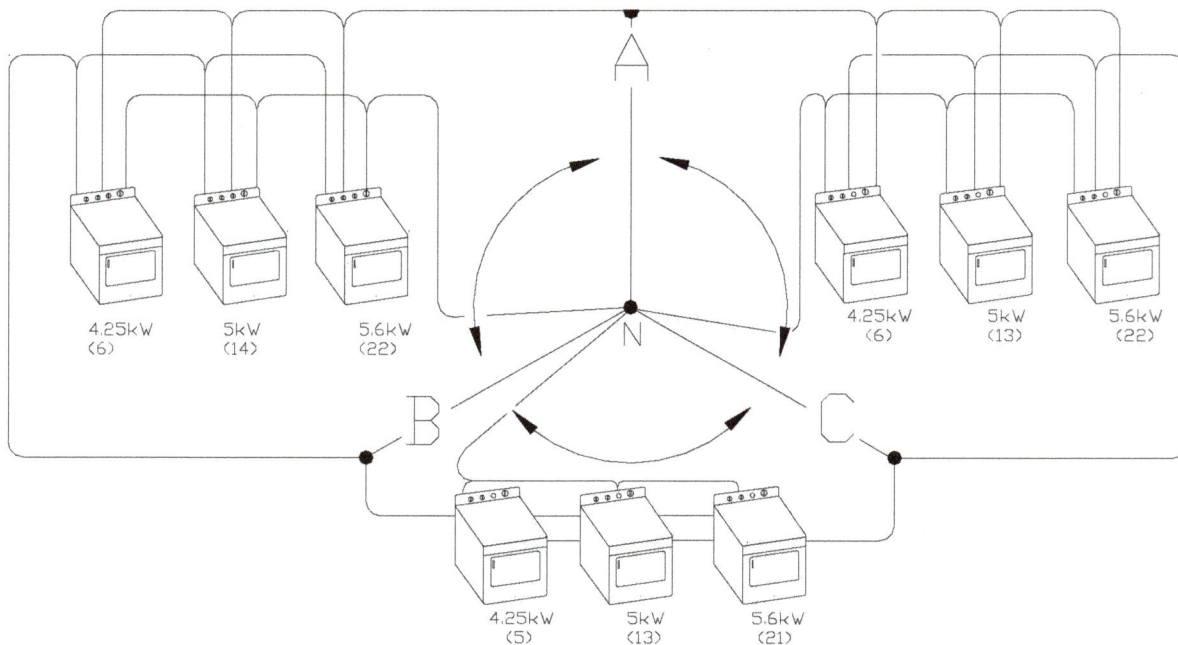

Figure 220.54-34

NEC 220.54 furthers states that, "the total load of the single-phase dryers must be computed on the basis of twice the maximum number of dryers connected between any two phases" which is accomplished by multiplying the number of dryers by 2.

6 dryers x 2 = **12 dryers**   14 dryers x 2 = **28 dryers**   22 dryers x 2 = **44 dryers**

This calculated result represents twice the maximum number of dryers connected between two phases.

The next step is to use the calculated results to determine the connected loads of the dryers and to derive the demand factors per **Table 220.54**. The connected loads of the dryers are,

$$\text{4250 } 5000 \text{ watts x } 12 = 60{,}000 \text{ watts}$$
$$5000 \text{ watts x } 28 = 140{,}000 \text{ watts}$$
$$5600 \text{ watts x } 44 = 246{,}400 \text{ watts}$$

As listed in Table 220.54 the demand factors for the 12, 28 and 44 dryers must be determined based on a specific demand percentage or given formula therefore,

$$\text{For } \textbf{12 dryers} \text{ - Demand Factor} = 47\% - [1\% \times (12 - 11)]$$
$$= 47\% - 1\% \times (1)$$
$$= 47\% - 1\%$$
$$= 46\% \text{ or } .46$$

The demand load per two phases is **27,600 watts** *where* 60,000 watts x .46 = **27,600 watts**.

$$\text{For } \textbf{28 dryers} \text{ - Demand Factor} = 35\% - .5\% \times (28 - 23)$$
$$= 35\% - .5\% \times (5)$$
$$= 35\% - 2.5\%$$
$$= 32.5\% \text{ or } .325$$

The demand load per two phases is **45,500 watts** *where* 140,000 watts x .325 = **45,500 watts**.

$$\text{For } \textbf{44 dryers} \text{ - Demand Factor} = 25 \text{ percent } (.25)$$

The demand load per two phases is **61,600 watts** *where* 246,400 watts x .25 = **61,600 watts**.

To determine the service demand load per phase (single-phase) for each set of dryers the calculated results are divided by 2.

$$27{,}600 \text{ watts} / 2 = 13{,}800 \text{ watts}$$
$$45{,}500 \text{ watts} / 2 = 22{,}750 \text{ watts}$$
$$61{,}600 \text{ watts} / 2 = 30{,}800 \text{ watts}$$

The service demand load for each unit of dryers (17, 40 and 65) supplied from a 3-phase, 4-wire system can now be determined by multiplying the single-phase loads by 3.

13,800 watts x 3 = **41,400 watts**   22,750 watts x 3 = **68,250 watts**
30,800 watts x 3 = **92,400 watts**

With these calculated results the 3-phase service demand load in amperes can now be determined for each unit by applying both single- and three-phase formulas.

	1φ		**3φ**
17 units	$\dfrac{13,800 \text{ watts}}{120\text{V}} = 115\text{A}$	or	$\dfrac{41,400 \text{ watts}}{208\text{V} \times 1.732} = 114.92 \text{ amps}$
40 units	$\dfrac{22,750 \text{ watts}}{120\text{V}} = 189.58\text{A}$	or	$\dfrac{68,250 \text{ watts}}{208\text{V} \times 1.732} = 189.45 \text{ amps}$
65 units	$\dfrac{30,800 \text{ watts}}{120\text{V}} = 256.67\text{A}$	or	$\dfrac{92,400 \text{ watts}}{208\text{V} \times 1.732} = 256.48 \text{ amps}$

220.55 and **Table** - Electric Ranges and Other Cooking Appliances - Dwelling Unit(s)

NEC 220.18(C) permits the application of the demand factors of Table 220.55 to include **Note 4** for household electric range loads.

In accordance with NEC 220.55, kilovolt-amperes (kVA) [or volt-amperes (VA)] shall be considered equivalent to kilowatts (kW) [or watts (W)] for loads calculated in this section.

Refer to **(Worksheet F** - Volume 4**)** DEMAND LOAD CALCULATIONS FOR HOUSEHOLD ELECTRIC RANGES AND OTHER COOKING APPLIANCES for related questions.

35. Explain the meaning of the statement listed in the heading of Table 220.55 which states in parenthesis "Column C to be used in all cases except as otherwise permitted in **Note 3**".

The values listed under Column C are the actual (maximum) demand loads based on the number of appliances referenced and require no calculations other than determining the maximum demand loads per given formulas where the number of appliances exceeds 25. Because Column C covers all appliances rated up to 12kW; those appliances with ratings that are applicable to Columns A and B falls within the boundaries of Column C also. Therefore, all calculated demand loads derived from Columns A or B should be compared to the demand loads in Column C to determine the lowest (minimum) demand load to be used. Such use is permitted at the conclusion (parenthesis) of the heading in Table 220.55 which implicitly permits the use of the lowest demand load.

Table 220.55, **Note 3** is the only provision aligned with Table 220.55 that softens the use of Column C because of the possibility that a lower demand load can be calculated where multiple cooking appliances exist among Columns A or B or both (cooking appliances over 1¾kW through 8¾kW). Often this consideration is ignored because it is assumed to be more logical or convenient to go exclusively to Column C for the demand load instead. In any case, the combined results (Columns A and B) obtained when applying **Note 3** must still be compared to Column C to determine the lowest (minimum) applicable demand load. See question Nos. 57. - 59. [Table 220.55, **Note 3** - Ranges rated over 1 ¾(1.75)kW through 8 ¾(8.75)kW].

Table 220.55, Column A - Household Electric Ranges,.......................................and Other Household Cooking Appliances (Less than 3½ kW) from 1.76kW to 3.49kW (36.-38.) [3]

The provisions of **Column A** are applied when a single appliance or a group of appliances having the same nameplate ratings are involved. Where household appliances consisting of different nameplate ratings are involved per **Column A**, the provisions of **Table 220.55, Note 3** is permitted. **Column A** covers household cooking appliances extending from 1.76kW to 3.49kW.

36. Determine the demand load for an electric cooktop that's supposedly rated for 1.7kW.

Because the appliance is rated less than the minimum 1.75kW rating per NEC 220.55, the demand load must be based on the actual rating (100 percent) of the appliance (1.7kW).

37. Determine the demand load for a cooktop rated for 3.25kW.

The demand load for an appliance rated less than 3½kW is determined by applying the demand factor given in Column A for one (1) appliance which is 80 percent (.80). Therefore,

$$3.25kW \times .80 = 2.6kW$$

Compared to the Maximum Demand of Column C (8kW) - Use 2.6kW as the demand load for the cooktop.

38. A cooktop appliance rated for 3.44kW is installed in each unit of a 17 unit apartment complex. What is the demand load of the appliances?

The total connected load of the appliances must be determined first which is,

$$3.44kW \times 17 = 58.48kW$$

The 38 percent (.38) demand factor given in Column A for 17 appliances can now be applied to determine the demand load of the appliances.

$$58.48kW \times .38 = 22.22kW$$

Compared to the Maximum Demand of Column C (32kW) - Use 22.22kW as the demand load for the cooktop appliances.

Table 220.55, Column B - Household Electric Ranges,.......................................and Other Household Cooking Appliances from 3½ (3.5) kW to 8 ¾ (8.75) kW (39. - 41.) [3]

The provisions of **Column B** are applied when a single appliance or a group of appliances having the same nameplate ratings are involved. Where household appliances consisting of different nameplate ratings are involved per **Column B**, the provisions of **Table 220.55, Note 3** is then applied. **Column B** covers household cooking appliances extending from 3.5kW to 8.75kW.

39. Determine the demand load for a 8.75kW household range.

The demand load for an appliance rated 8¾kW is determined by applying the demand factor given in Column B for one (1) appliance which is 80 percent (.80).

$$8.75kW \times .80 = 7kW$$

Compared to the Maximum Demand of Column C (8kW) - Use 7kW as the demand load for the range.

40. Determine the demand load for 14-5.4kW household ranges.

The total connected load of the household ranges is,

$$5.4kW \times 14 = 75.6kW$$

Applying the demand factor [32 percent (.32)] in Column B for 14 ranges

$$75.6kW \times .32 = 24.19kW$$

Compared to the Maximum Demand of Column C (29kW) - Use 24.19kW as the demand load for the household ranges.

41. What is the demand load for 22-7.5kW household ranges?

The total connected load of the appliances must be determined first which is,

$$7.5kW \times 22 = 165kW$$

The 26 percent (.26) demand factor given in Column B for 22 appliances can now be applied to determine the demand load of the appliances.

$$165kW \times .26 = 42.9kW$$

Compared to the Maximum Demand of Column C (37kW) - Use the maximum demand of Column C as the demand load for the 22 ranges because it has the lowest demand.

Table 220.55, Column C - Household Electric Ranges,..............................and Other Household Cooking Appliances (over 8¾kW up to 12kW) from 8.76kW to 12kW (42.-44.) [3]

42. What is the feeder demand load for 3-10.5 kW wall-mounted ovens?

Unlike the requirements used in Columns A and B where a demand factor (percent) has to be applied to the connected load to determine the demand load, the values given in Column C are either the actual demands or the actual demands derived per given formulas. Therefore, the feeder demand load for the 3 wall-mounted ovens per Column C is 14kW.

43. If 28 multifamily dwelling units are equipped with 10.5kW electric household ranges, determine the service demand load based on the given number of ranges.

The formula given in Column C for 28 appliances (26 - 30 appliances) must be applied to determine the service demand load. The formula requires 15kW to be added to the number of ranges based on a value of 1kW per range (where 28 ranges is equal to 28kW). As a result, 15kW + 28kW = 43kW.

The service demand load for the 28-10.5kW electric household ranges is 43kW opposed to the connected load, 294kW (10.5kW x 28).

44. If 128 multifamily dwelling units are equipped with 12kW electric household ranges, determine the service demand load based on the given number of ranges.

The formula given in Column C for 128 appliances (61 appliances and over) requires 25kW to be added to the number of ranges based on a value of ¾ (.75)kW per range. As a result,

$$25kW + (.75kW \times 128) = 121kW$$

The service demand load for the 128-12kW electric household ranges is 121kW opposed to the connected load, 1536kW (128 x 12kW).

Table 220.55, Note 1 - Ranges all of the same rating rated over 12kW through 27kW
(45. - 49.) [5]

45. Explain the term *major fraction*.

When the rating of an appliance includes a fractional (decimal) value that is either equal to or greater than .5 the equivalent or exceeding value defines a major fraction. When this occurs the rating of the appliance is allowed to be rounded up to the next whole number to compensate for the major fraction. For example, a group of household ranges are rated for 13.5kW, 13.7kW and 13.8kW. In this case the kilowatt rating of all ranges is rounded up to 14kW.

On the other hand when an appliance has a kilowatt rating consisting of a fractional value that does not include a major fraction, the kilowatt rating of the appliance is then rounded down to the preceding whole number. For example, two household ranges are rated for 13.2kW and 13.4kW. In this case the kilowatt rating of both ranges is rounded down to 13kW.

All in all, just remember fractional kilowatt ratings are only allowed to be used when applying the demands of Column C per Table 220.55, **Notes 1** and **2**.

In comparison to NEC 220.5(B) where the major fraction of an ampere is considered, the major fraction mentioned in **Notes 1** and **2** applies to the rating of an individual household range and not a calculated value.

46. Calculate the maximum demand load for a 17.5kW household electric range.

Although Table 220.55, **Note 1** specify the term "ranges", the same requirements are also followed for one electric range or any electric household cooking appliance(s). The maximum demand in Column C for one appliance must to be increased by 5 percent (.05) for each kilowatt exceeding 12kW up to 27kW.

To determine the demand load for the range, the following steps are followed:

1. Determine the kilowatts exceeding 12kW.

Since the kilowatt rating of the range (17.5kW) involves a major fraction the rating of the range is allowed to be rounded up to 18kW. Therefore,

$$18kW - 12kW = 6kW$$

2. Increase the number of kilowatts exceeding 12kW by 5 percent.

$$6 \times .05 = .30 + 1 = 1.30$$

The number of kilowatts exceeding 12kW is calculated at 30 percent (.30). Notice how the percentage value is added to 1 to derive the value 1.30. This value will be referred to as the *percent multiplier* and will be used to offer a short-cut. Observe how the *percent multiplier* is used in the next step.

3. Apply the percentage value to the maximum demand in Column C.

The maximum demand for 1 range per Column C is 8kW. This demand is increase by .30.

$$8kW \times .30 = 2.4kW$$

This value is now added to the 8kW maximum demand to reflect the 30 percent increase resulting in the maximum demand load.

$$8kW + 2.4kW = 10.4kW$$

By using the 1.30 *percent multiplier* the same results can be derived in one step that is,

$$8kW \times 1.30 = 10.4kW$$

This is what was meant earlier when stated that the *percent multiplier* offered a short-cut. Notice how the maximum demand (8kW) [1(100 percent)] and the percent increase (2.4kW) [(.30)] are included in the *percent multiplier* to render the same results as above. This short-cut approach will be utilized for all remaining questions where applicable.

47. Calculate the maximum demand load for a 27kW household electric range.

To determine the demand load for the range, the same steps are followed as in question No. 46.

1. Determine the kilowatts exceeding 12kW.

$$27kW - 12kW = 15kW$$

2. Increase the number of kilowatts exceeding 12kW by 5 percent.

$$15 \times .05 = .75 + 1 = 1.75$$

The number of kilowatts exceeding 12kW is increased by 75 percent (.75) resulting in a percent multiplier of 1.75.

3. Apply the percent multiplier to the maximum demand in Column C.

The maximum demand for 1 range per Column C is again 8kW. Therefore, the maximum demand load for the 27kW range is 14kW, (8kW × 1.75 = 14kW).

48. Determine the feeder demand load for 23-15kW household electric ranges.

The same steps are followed to determine the feeder demand load for the 23-15kW ranges.

1. Determine the kilowatts exceeding 12kW.

$$15kW - 12kW = 3kW$$

2. Increase the number of kilowatts exceeding 12kW by 5 percent.

$$3 \times .05 = .15 + 1 = 1.15$$

The number of kilowatts exceeding 12kW is increased by 15 percent (.15) resulting in a percent multiplier of 1.15.

3. Apply the percent multiplier to the maximum demand in Column C.

The maximum demand for 23 ranges per Column C is 38kW. Therefore, the feeder demand load for the 23 ranges is 43.7kW.

$$38kW \times 1.15 = 43.7kW$$

49. Determine the service demand load for 67-24.3kW household cooking appliances.

The same steps are followed to determine the service demand load for the 67-24.3kW household cooking appliances. Since the kilowatt rating of the appliances involves a fractional value less than .5, the appliances kilowatt rating is reduced to 24kW.

1. Determine the kilowatts exceeding 12kW.

$$24kW - 12kW = 12kW$$

2. Increase the number of kilowatts exceeding 12kW by 5 percent.

$$12 \times .05 = .60 + 1 = 1.60$$

The number of kilowatts exceeding 12kW is increased by 60 percent (.60) resulting in a percent multiplier of 1.60.

3. Apply the percent multiplier to the maximum demand in Column C.

The maximum demand for 67 ranges per Column C must be derived using the listed formula.

$$\text{Maximum demand} = 25kW + (.75kW \times 67) = 75.25kW$$

The service demand load for the 67 appliances can now be computed.

$$75.25kW \times 1.60 = 120.4kW$$

Compared to the appliance's connected load (67 x 24.3kW =1.6281MW [mega-watts) the service demand load (120.4kW) is far less due to the provisions of **Note 1**.

Table 220.55, Note 2 - Ranges of unequal ratings rated over 8 ¾(8.75)kW through 27kW (50. - 56.) [7]

50. Determine the average value of a 14kW, 15kW and 16kW range.

When two or more electric ranges or any electric household cooking appliances are unequally rated (individually) and the ratings exceeds 8¾kW up to 27kW Table 220.55, **Note 2** requires the average value of the ranges (appliances) to be computed by adding together all ranges to obtain the total connected load.

Based on the given ratings of the ranges, the total connected load is,

$$14kW + 15kW + 16kW = 45kW$$

Because there are 3 ranges the average value of the ranges is determined by dividing the connected load by the number of ranges where, 45kW / 3 = 15kW.

51. Determine the average value of a 9kW, 10kW, 13kW and 15kW range.

When a range (or cooking appliance) is rated over 8¾kW and less than 12kW, **Note 2** requires the rating of a range that's rated less than 12kW to be increased to 12kW before determining the average value of the ranges that are unequally rated.

Being the case, the ratings of the 9kW and 10kW ranges are individually increased to 12kW and the connected load is then computed.

$$(\cancel{9kW})12kW + (\cancel{10kW})12kW + 13kW + 15kW = 52kW$$

The average value of the ranges is,

$$52kW / 4 = 13kW$$

52. Determine the average value of an 11kW, 14kW, 18kW, 23kW and 27kW range.

The rating of the 11kW range is increased to 12kW before determining the connected load.

$$(\cancel{11kW})12kW + 14kW + 18kW + 23kW + 27kW = 94kW$$

The average value of the ranges is,

$$94kW / 5 = 18.8kW$$

Where the rating of an appliance involves a fractional (decimal) value, the major fraction must also be considered in **Note 2**.

Because the average value results in a major fraction that is greater than .5 (.8), the value is required to be rounded up to 19kW.

53. Determine the average value of a 13kW, 15.6kW, 16kW, 21.5kW, 24.4kW and 26.3kW range.

The connected load is,

$$13kW + 15.6kW + 16kW + 21.5kW + 24.4kW + 26.3kW = 116.8kW$$

The average value of the ranges is,

$$116.8kW / 6 = 19.4667kW$$

Because the average value results in a fractional value that is less than .5 (.4667), the value is required to be rounded down to 19kW.

54. Determine the demand load for the ranges given in question No. 50.

Once an average value has been obtained per **Note 2**, the requirements are the same as used for **Note 1** which requires the maximum demand in Column C per number of appliances to be increased by 5 percent (.05) for each kilowatt exceeding 12kW up to 27kW.

Using the average value (15kW) of the ranges, the same steps can be used as applied with **Note 1** to determine the demand load of the ranges.

1. Determine the kilowatts exceeding 12kW.

$$15kW - 12kW = 3kW$$

2. Increase the number of kilowatts exceeding 12kW by 5 percent.

$$3 \text{ x } .05 = .15 + 1 = 1.15$$

The number of kilowatts exceeding 12kW is increased by 15 percent (.15) resulting in a percent multiplier of 1.15.

3. Apply the percent multiplier to the maximum demand in Column C.

The maximum demand for 3 ranges per Column C is 14kW. Therefore, the demand load is,

$$14kW \text{ x } 1.15 = 16.1kW$$

55. Determine the feeder demand load for 2-9.5kW ranges, 3-14.8kW ranges, 2-16.6kW ranges and 5-19.4kW ranges.

Considering the fact that there are ranges with the same ratings, those ranges having the same rating are not to be evaluated separately per **Note 1** and then added together. Calculating the feeder demand load of the ranges requires being performed as a group and for that reason **Note 2** is applied.

1. Determine the average value of the ranges.

The rating of the 9.5kW ranges is increased to 12kW before determining the connected load.

$$
\begin{array}{lllll}
12.0kW \text{ (\sout{9.5kW})} & \text{ranges} & \text{x} & 2 = & 24.0kW \\
14.8kW & \text{ranges} & \text{x} & 3 = & 44.4kW \\
16.6kW & \text{ranges} & \text{x} & 2 = & 33.2kW \\
19.4kW & \text{ranges} & \text{x} & \underline{5} = & \underline{97.0kW} \\
& & & 12 & 198.6kW
\end{array}
$$

$$198.6kW / 12 = 16.55kW$$

Because the average value results in a major fraction that is greater than .5 (.55), the value is required to be rounded up to 17kW.

2. Determine the kilowatts exceeding 12kW.

$$17kW - 12kW = 5kW$$

3. Increase the number of kilowatts exceeding 12kW by 5 percent.

$$5 \times .05 = .25 + 1 = 1.25$$

The number of kilowatts exceeding 12kW is increased by 25 percent (.25) resulting in a percent multiplier of 1.25.

4. Apply the percent multiplier to the maximum demand in Column C.

The maximum demand for 12 ranges per Column C is 27kW. Therefore, the feeder demand load is,

$$27kW \times 1.25 = 33.75kW$$

56. A 75 unit condominium complex is equipped with either electric ranges or other household cooking appliances. In 23 units - 9.2kW cooktops and 9kW wall-mounted ovens (wmo) are installed, in 32 units - 17.6kW ranges are installed and in 20 units - 9.2kW cooktops and 10.4kW microwave (m-w)/oven combinations are installed. Determine the service demand load based on the installed cooking appliances.

1. Determine the average value of the ranges.

12.0kW(~~9.0kW~~) wmo	x 23 =	276.0kW
12.0kW(~~9.2kW~~) cooktops	x 43 =	516.0kW
12.0kW(~~10.4kW~~) m-w/ovens	x 20 =	240.0kW
17.6kW ranges	x <u>32</u> =	<u>563.2kW</u>
	118	1595.2kW* (connected load)

*(or 1.5952MW [mega-watts])

$$1595.2kW / 118 = 13.52kW$$

Because the average value results in a major fraction that is greater than .5 (.52), the value is required to be rounded up to 14kW.

2. Determine the kilowatts exceeding 12kW.

$$14kW - 12kW = 2kW$$

3. Increase the number of kilowatts exceeding 12kW by 5 percent.

$$2 \times .05 = .10 + 1 = 1.10$$

The number of kilowatts exceeding 12kW is increased by 10 percent (.10) resulting in a percent multiplier of 1.10.

4. Apply the percent multiplier to the maximum demand in Column C.

The maximum demand for 118 appliances per Column C must be derived using the listed formula.

$$\text{Maximum demand} = 25kW + (.75kW \times 118) = 113.5kW$$

The service demand load for the 118 appliances is,

$$113.5kW \times 1.10 = 124.85kW$$

Although the connected load is approximately 12.78 times (1595.2kW/124.85kW) greater than the service demand load, only 7.83 percent (.0783) of the connected load 124.85kW/1595.2kW) is used towards sizing the service and service conductors for this application per calculated service demand load. Because such massive derating applications can cause doubt and perhaps a great deal of uncertainty; the application and final results are both practical and code compliant.

Table 220.55, Note 3 - Ranges rated over 1 ¾(1.75)kW through 8 ¾(8.75)kW (57. - 59.) [3]

57. Determine the demand load for 8-3kW cooktops and 8-5.8kW ranges.

The demand load for electric household cooking appliances, rated over 1¾kW up to 8¾kW, could in fact be determined based on the maximum demand values listed under Column C of Table 220.55, since Column C is inclusive of all appliances rated over 1¾kW up to 12kW. However, just remember where the provisions of Table 220.55, **Note 3** are applied the use of Column C only serves as a comparison factor to determine the lowest demand load. See question No. 35.

Table 220.55, **Note 3** permits an alternative method for determining the demand load for appliances rated over 1¾kW up to 8¾kW by: **(1)** allowing the nameplate ratings of all rated appliances to be added together and multiplied by the applicable demand factors given in either Column A or B per given number of appliances and **(2)** when the nameplate rating of a cooking appliance falls under both Columns A and B, the applicable demand factors per column is applied to the nameplate rating for that appliance and the results gathered from both columns are added together.

Since the nameplate ratings of the appliances listed in the question fall under both Columns A and B the demand factors will be applied accordingly and the results derived from the use of both columns combined. The comparison of Column C is based upon the combined results.

Column A

$$3kW \times 8 = 24kW$$
$$(8 \text{ appliances} = .53 \text{ demand factor})$$
$$24kW \times .53 = \mathbf{12.72kW}$$

Column B

$$5.8kW \times 8 = 46.4kW$$
$$(8 \text{ appliances} = .36 \text{ demand factor})$$
$$46.4kW \times .36 = \mathbf{16.7kW}$$

Results of **Columns A** and **B**

$$12.72kW + 16.7kW = \mathbf{29.42kW}$$

The demand load for the 8 cooktops and 8 ranges as totaled per Columns **A** and **B** is **29.42kW**. Compared to Column C, the maximum demand for 16 appliances is 31kW. Therefore, the results per Columns A and B are used which is the lowest demand load.

58. Determine the demand load for 6-2.25kW cooktops, 4-3.45kW cooktops, 7-7.6kW wall-mounted ovens and 5-8.7kW ranges.

Again, since the nameplate ratings of the appliances fall under both Columns A and B the demand factors will be applied accordingly and the results derived from the use of both columns combined.

Column A

$$2.25kW \times 6 = 13.5kW$$
$$3.45kW \times \underline{4} = \underline{13.8kW}$$
$$ 10 \quad 27.3kW$$
$$(10 \text{ appliances} = .49 \text{ demand factor})$$
$$27.3kW \times .49 = \mathbf{13.38kW}$$

Column **B**

$$7.6kW \times 7 = 53.2kW$$
$$8.7kW \times \underline{5} = \underline{43.5kW}$$
$$ 12 \quad 96.7kW$$
$$(12 \text{ appliances} = .32 \text{ demand factor})$$
$$96.7kW \times .32 = \mathbf{30.94kW}$$

Results of Columns **A** and **B**

$$13.38kW + 30.94kW = \mathbf{44.32kW}$$

The demand load for the appliances is 44.32kW. In this case, the maximum demand of Column C is used, **37kW** based on 22 appliances.

59. Determine the service demand load for 10-3.3kW cooktops, 16-8.5kW wall-mounted ovens and 19-11.45kW ranges.

This question may appear to require partial use of the application of **Note 3**. However, if you take a closer look at the ratings of all involved appliances you will notice that all appliances are rated less than 12kW. As a result, the service demand for all appliances (45) can be derived directly from Column C without having to consider each column individually. In other words, for an application such as this the use of **Note 3** is not practical. To prove the impracticality of this matter we will approach the question applying the provisions of **Note 3** and then Column C. Observe.

$$\text{Per } \textbf{Note 3} - 3.3\text{kW x } 10 \text{ x } .49 = 16.17\text{kW}$$
$$8.5\text{kW x } 16 \text{ x } .28 = \underline{38.08\text{kW}}$$
$$54.25\text{kW}$$

$$\text{Column C (19 ranges) max. demand} = \underline{34.00\text{kW}}$$
$$\textbf{88.25kW} \text{ (Total \textbf{Note 3} and Col C)}$$

The demand factors in Column **A** and **B** were used to determine the demand loads for each respective group of appliances per **Note 3** followed by the use of Column C resulting in a service demand load of **88.25kW**.

Now, considering all 45 appliances collectively, per Column C the maximum demand is,

$$25\text{kW} + (.75\text{kW x } 45) = \textbf{58.75kW}$$

As you can see if the methods used in **Note 3** and Column C were applied together the results would be far greater than the given service demand load per Column C thus rending such application impractical.

Table 220.55, Notes 1 and **3 - Combinations** - Columns (A, B and C)

60. Determine the service demand load for the appliances in question No. 59. if the ranges are rated for 21kW instead.

Using ranges that are rated higher than 12kW will certainly cause a different application opposed to the application used in question No. 59. For sure the total number of appliances cannot be considered collectively.

The demand load for the 21kW ranges must be determined per **Note 1** whereas the demand load for the 3.3kW and 8.5kW ranges can be determined per **Note 3**.

Applying the conditions of **Note 1** to determine the demand load for 19-21kW ranges.

1. Determine the kilowatts exceeding 12kW.

$$21kW - 12kW = 9kW$$

2. Increase the number of kilowatts exceeding 12kW by 5 percent.

$$9 \text{ x } .05 = .45 + 1 = 1.45$$

The number of kilowatts exceeding 12kW is increased by 45 percent (.45) resulting in a percent multiplier of 1.45.

3. Apply the percent multiplier to the maximum demand in Column C.

The maximum demand for 19 ranges per Column C is 34kW. Therefore, the demand load for the 19 ranges is,

$$34kW \text{ x } 1.45 = \textbf{49.3kW}$$

Applying the conditions of **Note 3** to determine the demand load for the 3.3kW and 8.5kW ranges.

Column A

$$3.3kW \quad x \quad 10 = 33kW$$
$$(10 \text{ appliances} = .49 \text{ demand factor})$$
$$33kW \text{ x } .49 = 16.17kW$$

Column B

$$8.5kW \quad x \quad 16 = 136kW$$
$$(16 \text{ appliances} = .28 \text{ demand factor})$$
$$136kW \text{ x } .28 = 38.08kW$$

The results per Column **A** and **B** is **54.25kW** (16.17kW + 38.08kW = **54.25kW**) whereas, the results per Column C based on 26 appliances is **41kW** (15kW + 26kW = **41kW**). In conclusion, the results of Column C are added to the demand load of the 19-21kW ranges to derive the service demand load.

The service demand load for this application is **90.3kW** (41kW + 49.3kW = **90.3kW**). Because the ranges are rated over 12kW they cannot be grouped with the cooktops and wall-mounted ovens where the maximum demand for all appliances could be determined per Column C as performed in question No. 59.

Table 220.55, Note 4 - Branch-Circuit Load (61. - 65.) [5]

61. Determine: **(1)** The branch-circuit loads, **(2)** The size branch-circuit conductors (using nonmetallic-sheathed cable [Romex]) and **(3)** the overcurrent devices required to protect the ranges in question Nos. 39. and 46. Assume the voltage supplying the appliances is rated for 240/120V.

For an individual range, Table 220.55, **Note 4** permits the maximum demand listed in Column C to be used for determining the branch-circuit load for one range. However, to determine the branch-circuit load for an individual wall-mounted oven or counter-mounted cooking unit (cooktop), Table 220.55, **Note 4** requires the nameplate rating of such appliances to be used opposed to a calculated demand load *or* maximum demand per Table 220.55.

(1) COMPUTING BRANCH-CIRCUIT LOADS

In question No. 39. the demand load for the 8.75kW range was found to be 7kW after using the demand factor given for one appliance in Column B. As a result, the branch-circuit load for the range is 7kW / 240V = **29.17A**.

In question No. 46. the demand load for the 17.5kW range was found to be 10.4kW after increasing the maximum demand in Column C (for one range) according to **Note 1**. As a result, the branch-circuit load for the range is 10.4kW / 240V = **43.3A**.

Assuming a 240/120V source, the 8.75kW range has a rated current of 36.46A (8.75kW / 240V) and the 17.5kW range has a rated current of 72.92A (17.5kW / 240V). **Note 4** permits the branch-circuit load for both ranges to be computed based on the provisions of Table 220.55 opposed to using each range rated current. This is because it can easily be assumed that the ranges will almost never operate at their rated current. Therefore, the branch-circuit load of the 8.75kW range is permitted to be calculated at 80 percent (.80) (29.17A / 36.46A) of its rated current while the branch-circuit load of the17.5kW range is permitted to be calculated at 59.4 percent (.594) (43.3A / 72.92A) of its rated current.

(2) DETERMINING BRANCH-CIRCUIT CONDUCTORS

NEC 422.10(A) states that the rating of an individual branch circuit supplying an appliance must not be less than the marked rating of the appliance, however in the concluding paragraph of that section it specifically states that branch circuits for household cooking appliances shall be permitted to be in accordance with Table 220.55 and shall be sized in accordance with NEC 210.19(A)(3).

According to NEC 334.80 the ampacity of Nonmetallic-Sheathed Cable (Romex) must not exceed that of a 60°C rated conductor. Referring to (60°C temperature rating for copper conductors) Table 310.15(B)(16), to select the appropriate size conductors for each range per calculated branch-circuit load, 10 AWG conductors (10-3 with equipment grounding conductor) which have an ampacity of 30 amps can be used for the 8.75kW range and 6 AWG conductors (6-3 with equipment grounding conductor) which have an ampacity of 55 amps can be used for the 17.5kW range.

Three conductor nonmetallic-sheathed cable with an equipment grounding conductor must be used according to NEC 250.134 and 250.138 for new branch circuits because the grounded (neutral) conductor cannot be used as an equipment grounding conductor for ranges and clothes dryers per NEC 250.140. A separate grounded (neutral) conductor and equipment grounding conductor are required.

(3) DETERMINING BRANCH-CIRCUIT OVERCURRENT DEVICES

NEC 422.11(A) references the use of NEC 240.4 for determining the branch-circuit overcurrent protection for appliances. As permitted, NEC 240.4(D)(7) requires for this application the maximum use of a 30 amps overcurrent device based on the use of 10 AWG copper conductors for the 8.75kW range whereas, NEC 240.4(B) allows the use of a 60 amps overcurrent device based on the use of 6 AWG copper conductors for the 17.5kW range.

On the other hand, NEC 210.19(A)(3) states that ranges rated for 8¾(8.75)kW or more are required to have a minimum branch-circuit rating of 40 amps. Because of this requirement, the branch-circuit rating (overcurrent device) for the 8.75kW range has to be increased to 40 amps and 8-3 nonmetallic-sheathed cable which has an ampacity of 40 amps is required to be used as the branch-circuit conductors for the range instead of 10-3. Just remember there is a difference between the *nameplate rating* of a range and the *branch-circuit load*.

62. Determine the branch-circuit load for a 6.8kW single wall-mounted oven and a 7.6kW double wall-mounted oven rated for 240/120V. What size individual branch-circuit conductors (60°C copper) and overcurrent devices are required to supply each oven?

Note 4 requires the branch-circuit load for *a* wall-mounted oven (whether single or double mounted) or *a* counter-mounted (cooktop) cooking unit to be computed based on the *nameplate rating* of the appliance. Because the branch-circuit load for one range is permitted to be computed per Table 220.55, the requirement for computing the branch-circuit load for a wall-mounted oven or counted-mounted cooking unit is often misinterpreted and assumed to be the same as that for a range.

The branch-circuit loads based on the *nameplate ratings* of the wall-mounted ovens are,

$$\frac{6800W}{240V} = 28.33A \qquad \frac{7600W}{240V} = 31.67A$$

Referring to Table 310.15(B)(16) (60°C copper), 10 AWG conductors are required for the 6.8kW oven and 8 AWG conductors are required for the 7.6kW oven. As for overcurrent devices, a 35A device must be used for the 7.6kW oven whereas NEC 240.4(D)(7) specifically requires as a maximum, a 30A device for the 6.8kW oven.

63. A 240V-6.3kW cooktop unit and a 240/120V-5.4kW wall-mounted oven are installed in the kitchen of a dwelling. Determine the branch-circuit loads for the appliances.

Instead of having to compute the branch-circuit load per nameplate rating for each appliance separately as in question No. 62., **Note 4** permits otherwise. When a counter-mounted cooking unit and no more than two wall-mounted ovens are located in the same room and supplied from a single branch circuit the nameplate ratings of the appliances are permitted to be added together and treated as one range according to **Note 4**.

Since the appliances can be supplied from the same circuit, the circuit's connected load and the appliance's total rating is 11.7kW (6.3kW + 5.4kW). If permitted by **Note 4** to be treated as one range, the branch-circuit load can be determined based on the demand load given in Column C of Table 220.55, because the total rating of the appliances when treated as one range does not exceed 12kW. Being the case, the maximum demand for one range is 8kW and the branch-circuit load for both appliances is calculated at 33.3A (8000W / 240V).

However, because the two appliances are treated as one range the provisions of NEC 210.19(A)(3) must still be applied meaning the minimum branch-circuit rating (overcurrent device) cannot be less than 40 amperes and so must the supplying cable (conductors).

64. (a) Determine the branch-circuit load if another 5.4kW wall-mounted oven is added to the circuit in question No. 63. What size branch-circuit overcurrent device, single branch-circuit conductors and tap conductors (60°C copper) are required to supply the appliances?

Adding another appliance increases the total rating of the appliances to 17.1kW (6.3kW + 5.4kW + 5.4kW), which is still permitted to be treated as one range per **Note 4**. Since the rating exceeds 12kW, **Note 1** to Table 220.55 is applied.

To determine the branch-circuit load for this application, the same steps are followed as previously. Since the kilowatt rating involves a fractional value less than .5, the rating is reduced to 17kW.

1. Determine the kilowatts exceeding 12kW.

$$17kW - 12kW = 5kW$$

2. Increase the number of kilowatts exceeding 12kW by 5 percent.

$$5 \times .05 = .25 + 1 = 1.25$$

The number of kilowatts exceeding 12kW is increased by 25 percent (.25) resulting in a percent multiplier of 1.25.

3. Apply the percent multiplier to the maximum demand in Column C.

The maximum demand for one range per Column C is 8kW. Therefore, the maximum demand load determines the branch-circuit load for this application which is 10kW (8kW x 1.25). The single branch circuit which will supply the cooktop and wall-mounted ovens is determined based on the calculated load,

$$10kW / 240V = 41.67A$$

Referring to Table 310.15(B)(16), 6 AWG conductors, which have an ampacity of 55 amps at 60°C are required. Although a 60 amps overcurrent device could be used [NEC 210.23(D) and 240.4(B)], NEC 210.19(A)(3), *Exception 1* provides specific requirements which permits the use of tap conductors when branch-circuit conductors supplying household ranges and cooking appliances are protected by a 50 amps overcurrent device, where the ampacity of the tap conductors is not less than 20 amps (minimum 12 AWG copper conductor) and sufficient to supply the load being served. *Exception 1* also requires the length of a tap conductor to be only as long as needed.

If installing a 6-3 nonmetallic-sheathed cable as the branch-circuit conductors and routing it to a junction box to supply the cooktop and two wall-mounted ovens, individual tap conductors can be used to supply each appliance. Calculating the load for each appliance per *nameplate rating* (**Note 4** to Table 220.55) to size the tap conductors results in the following,

Cooktop - 6300W / 240V = **26.3A**   **Ovens** - 5400W / 240V = **22.5A**

A 10-2 (line to line load) nonmetallic-sheathed cable (30 amps ampacity) can be used as the tap conductors to supply the cooktop and two 10-3 nonmetallic-sheathed cables (30 amps ampacity) can be used as tap conductors to supply the wall-mounted ovens. Both sets of cables have the capacity to supply each load and have ampacities greater than 20 amps. Refer to Figure 220.55, **Note 4**-64.

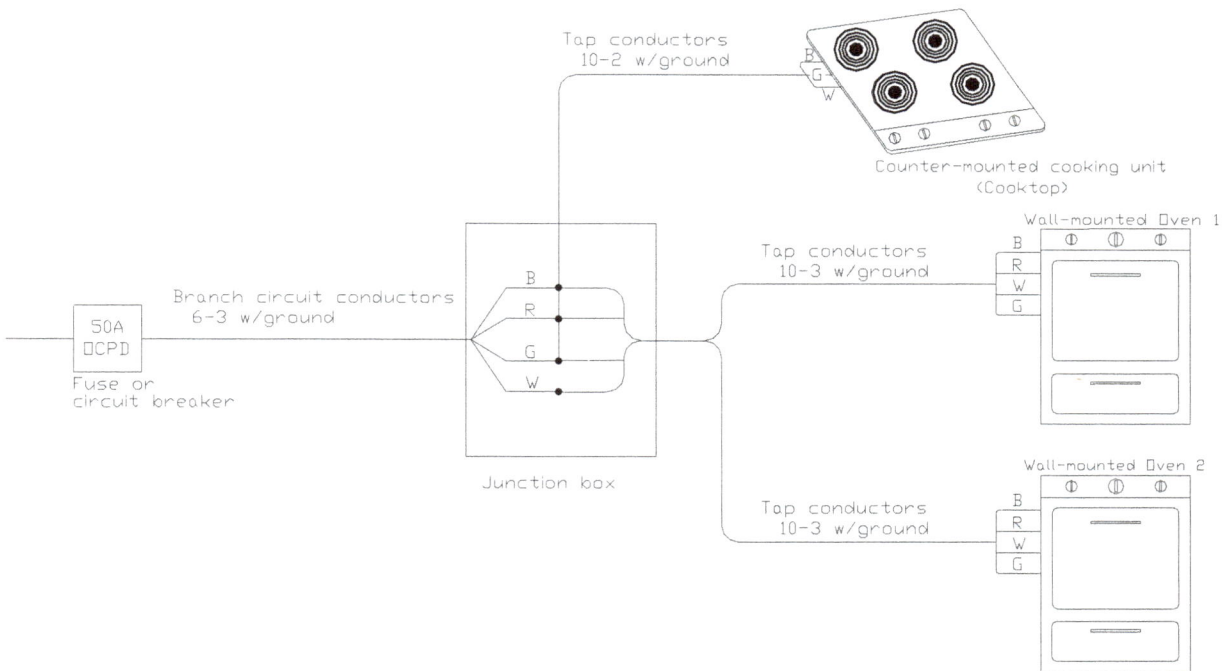

Figure 220.55, **Note 4**-64

(b) Now suppose the cooktop and wall-mounted ovens are replaced with a 2.4kW cooktop and dual wall-mounted ovens rated for 3.1kW. Determine the branch-circuit load, the size branch-circuit device; single branch-circuit conductors and tap conductors required to supply these appliances. All conductors to be rated for 60°C copper.

Just as the previous appliances, the ratings of these appliances are added together which amounts to 8.6kW (2.4kW + 3.1kW + 3.1kW) and is also permitted to be treated as one range per **Note 4**. Since the total rating does not exceed 8.75kW and not to mention 12kW, the provisions of Columns B and C can be applied and then compared to determine which application provides the lowest value.

In accordance with **Note 4,** based on being treated as one range, per **Column B** an 80 percent demand factor can be applied resulting to a demand load of 6.88kW (8.6kW x .80). On the other hand, per **Column C** one range yields an 8kW maximum demand. As a result, the demand load per **Column B** is used as the branch-circuit load leading to the branch-circuit overcurrent device and conductors being sized based upon the following calculated load current,

$$6.88kW / 240V = \textbf{28.67A}$$

At 28.67A, a 30A branch-circuit overcurrent device can be used along with 10 AWG copper branch-circuit conductors at 60°C per Table 310.15(B)(16) which has a rated ampacity exact to that of the overcurrent device.

To size the tap conductors supplying the cook-top and wall-mounted ovens, the nameplate ratings of the appliances per **Note 4** must be used where,

$$2.4kW / 240V = 10A \quad 3.1kW / 240V = 12.92A$$

Referring to Table 310.15(B)(16), 14 AWG copper conductors, which have an ampacity of 15A at 60°C can be used to supply each individual appliance. With the use of a 30A overcurrent device and 10 AWG conductors being used as the branch-circuit conductors per Table 210.24, 14 AWG conductors are permitted to be used as tap conductors which are the same size conductors needed to supply each appliance.

In comparison to question 64.(a) and Figure 220.55, **Note 4**-64, the resulting 30A overcurrent device and 10 AWG branch-circuit conductors will in this situation serve the same purpose as the 50A overcurrent device and 6-3 branch-circuit conductors while the 14 AWG tap conductors serves the same purpose as the 10-2 and 10-3 tap conductors supplying the cooktop and wall-mounted ovens . Also in this situation, the 10 and 14 AWG conductors would be used in the same combinations as those shown in Figure 220.55, **Note 4**-64.

In concluding, the intended purpose of Questions 64.(a) and (b) along with solutions is to further demonstrate the provisions of **Note 4** when such combinations are treated as one range rated *less* or *greater* than 12kW.

65. If three (3) TW copper conductors with an equipment ground were installed in a raceway and used as the branch-circuit conductors in question No. 64.(a) instead of a 6-3 nonmetallic-sheathed cable, what size neutral conductor could be used?

Where the maximum demand has been computed using Column C of Table 220.55, NEC 210.19(A)(3), *Exception 2* permits the ampacity of the neutral conductor for ranges rated 8¾kW or more to be smaller than the ungrounded conductors and reduced no less than 70 percent of the *branch-circuit rating* but never smaller than a 10 AWG conductor.

Again 6 AWG TW copper conductors will be used as the branch circuit's ungrounded conductors. With a 50 amps branch-circuit rating, the ampacity of the neutral conductor can be calculated at 70 percent.

$$50A \times .70 = 35A$$

Referring to Table 310.15(B)(16), an 8 AWG TW copper conductor which has an ampacity of 40 amps could be used as the branch-circuit's neutral conductor.

As for the size of the equipment grounding conductor, if needed, it must be sized according to Table 250.122 which requires a 10 AWG copper conductor based on the use of a 50 amps overcurrent device.

Table 220.55, Note 5 - Household cooking appliances rated over 1¾(1.75)kW used in Instructional Programs

66. The culinary arts department of a technical college is equipped with 12-10.6kW and 13-11.7kW electric household ranges. Determine the feeder demand load for these appliances.

Where household cooking appliances rated over 1¾kW are used in an instructional program such as a high school home economics classroom or the one referenced in the question, Table 220.55, **Note 5** permits the use of Table 220.55 for determining the demand loads.

Because the ranges are rated over 8¾kW and less than 12kW, the 40kW maximum demand listed in Column C for 25 appliances is the feeder demand load for these appliances.

Single-Phase Ranges on Three-Phase, 4-Wire System (67. - 70.) [4]

67. Determine the service demand load when 73-9.5kW ranges are fed from a 4-wire, 3-phase, 208/120V source.

According to NEC 220.55, where two or more single-phase ranges are supplied by a 3-phase, 4-wire service, the total load of the single-phase ranges must be computed on the basis of twice the maximum number of ranges connected between any two phases. This same procedure was used to compute the demand load for electric dryers in question No. 34. Now observe how it is used to compute the demand load for household electric ranges.

Again, the 3-phase, 4-wire service referenced in NEC 220.55 refers to a 3-phase, 4-wire, 208/120V wye-connected source. In a 3-phase, 4-wire, wye-connected system there are two phase windings between each line-to-line connection which means this occurs three times. So now let's determine the maximum number of ranges connected between any two phases by dividing the number of ranges by the number of phases,

$$\frac{73 \text{ ranges}}{3} = 24.3 \text{ (rounded up to nearest whole number)} = 25$$

Since 25 ranges is the maximum number that can be connected between any two phases, the remaining 48 ranges will be divided between the other sets of two-phase windings. See Figure 220.55-67.

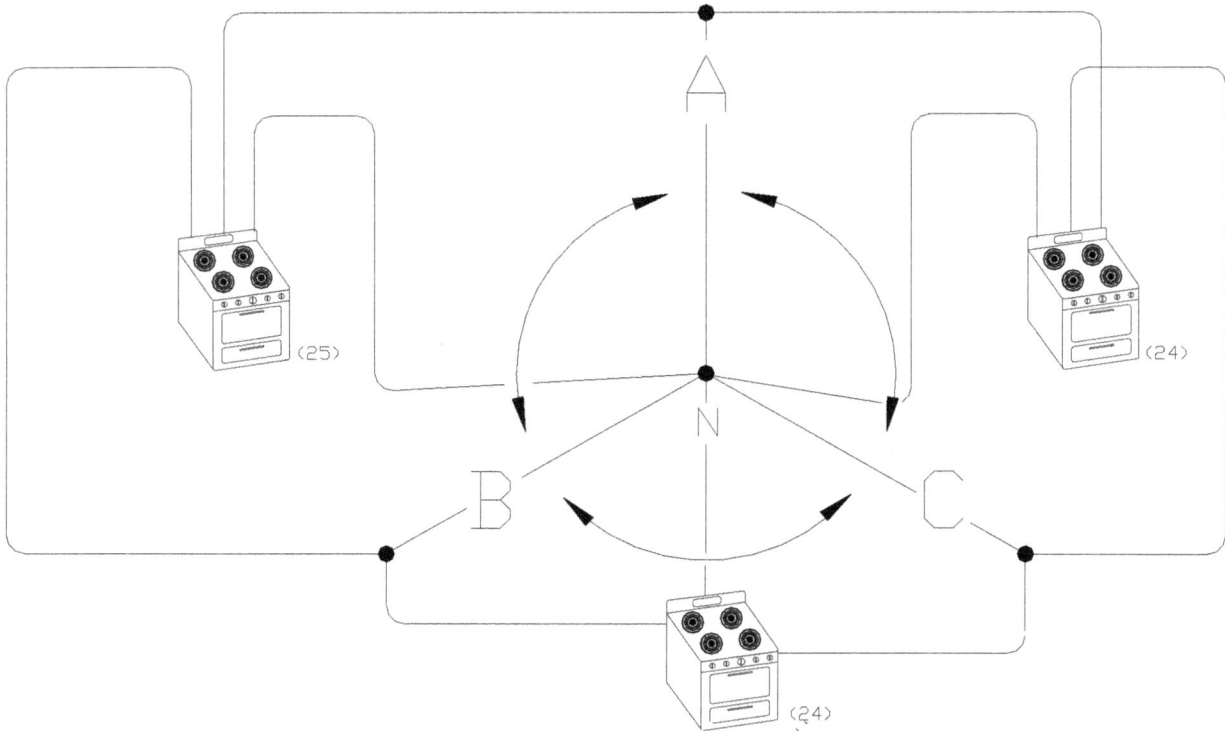

Figure 220.55-67

NEC 220.55 furthers states that, "the total load of the single-phase ranges must be computed on the basis of twice the maximum number of ranges connected between any two phases" which is accomplished by multiplying the 25 ranges by 2.

25 ranges x 2 = 50 ranges

This calculated value represents the total number of ranges between two phases.

The next step is to use the total number of ranges to determine the maximum demand per Table 220.55. As listed in Column C, the maximum demand factor for 50 ranges must be determined based on the given formula therefore,

(Maximum Demand) = 25kW + (.75kW x 50) = 62.5kW (62,500 watts)

The maximum demand (service demand load) per two phases is 62.5kW. To determine the service demand load per phase (single-phase) this value is divided by 2.

62,500 watts / 2 = 31,250 watts

With this value the single-phase service demand load in amps can be determined.

$$\frac{31,250 \text{ watts}}{120V} = 260.42 \text{ amps}$$

The service demand load of the 73 single-phase ranges supplied from a 3-phase, 4-wire system can now be determined by multiplying the single phase load by 3.

31,250 watts x 3 = **93,750 watts**

With this calculated value the 3-phase service demand load in amperes can now be determined which should be the same (approximate) value calculated using the single-phase demand load.

Three-phase Demand Load

$$\frac{93,750 \text{ watts}}{208V \times 1.732} = \textbf{260.23 amps}$$

68. Determine the service demand load when 43-6.35kW ranges are fed from a 4-wire, 3- phase, 208/120V source.

Using the abbreviated step by step method.

1. Divide the number of ranges by the number of phases.

$$\frac{43 \text{ ranges}}{3} = 14.3 \text{ (rounded up to nearest whole number)} = 15$$

2. Determine the maximum number of ranges connected between any two phases.

15 ranges x 2 = 30 ranges

3. Determine the demand load of the ranges per demand factor in Column B.

6.35kW x 30 x .24 = 45.72kW

45.72kW represents the service demand load per two phases.

Compare. Determine the demand load of the ranges per maximum demand in Column C. The maximum demand for 30 ranges is 45kW (15kW + 30kW) which is lower, although slightly, than the calculated results, 45.72kW.

4. Determine the single-phase (**1ϕ**) and three-phase (**3ϕ**) service demand loads.

$$\frac{45kW}{2} = \textbf{22.5kW} \text{ [service demand load per phase (1ϕ)]}$$

22.5kW x 3 = **67.5kW** [three-phase **(3ϕ)** service demand load]

5. Determine the service demand load in amperes (A).

$$\frac{22.5kW}{120V} = \textbf{187.5A} \text{ [per phase (1ϕ)]} \quad \text{-or-} \quad \frac{67.5kW}{208V \times 1.732} = \textbf{187.37A} \text{ [three-phase (3ϕ)]}$$

69. Determine the service demand load when 27-19kW ranges are fed from a 4-wire, 3- phase, 208/120V source.

Using the abbreviated step by step method.

1. Divide the number of ranges by the number of phases.

$$\frac{27 \text{ ranges}}{3} = 9$$

2. Determine the maximum number of ranges connected between any two phases.

9 ranges x 2 = 18 ranges

3. Determine the demand load of the ranges per **Note 1**.

Determine the kilowatts exceeding 12kW.

19kW – 12kW = 7kW

Increase the number of kilowatts exceeding 12kW by 5 percent.

7 x .05 = .35 + 1 = 1.35

Apply the percent multiplier to the maximum demand in Column C.

The maximum demand for 18 ranges per Column C is 33kW. Therefore, the service demand load per two phases is 33kW x 1.35 = 44.55kW.

4. Determine the single-phase (1ϕ) and three-phase (3ϕ) service demand loads.

$$\frac{44.55kW}{2} = \textbf{22.3kW} \text{ [service demand load per phase (1ϕ)]}$$

22.3kW x 3 = **66.9kW** [three-phase **(3ϕ)** service demand load]

5. Determine the service demand load in amperes (A).

$$\frac{22.3kW}{120V} = \textbf{185.83A} \text{ [per phase (1}\phi\text{)]} \quad \text{-or-} \quad \frac{66.9kW}{208V \times 1.732} = \textbf{185.7A} \text{ [three-phase (3}\phi\text{)]}$$

70. Determine the service demand load where the following cooking appliances are fed from a 4-wire, 3-phase, 208/120V source.

15 - 6.1kW ranges   17 - 7.8kW ranges
12 - 6.3kW cooktops   10 - 6.7kW cooktops
12 - 5.4kW WMOs   10 - 5.8kW WMOs

This selection of cooking appliances is also used in question No. 87.

Because NEC 220.55 only mentions <u>single-phase ranges</u> being supplied by a 3-phase, 4-wire feeder or service, the described method cannot be totally used in this situation based upon the given combination of ranges and other type cooking appliances. Therefore, this question will only apply the provisions of Columns B (3½kW to 8¾kW) and C (Not over 12kW) to derive the service demand load. Because steps 1. and 2. as previously used are not applicable this process will begin with step 3.

3. Determine the demand load of appliances per demand factor in Column B. The appliances total kW rating (connected load) must first be determined.

Ranges -	15 x 6100W =	91,500W
Ranges -	17 x 7800W =	132,600W
Cooktops -	12 x 6300W =	75,600W
Cooktops -	10 x 6700W =	67,000W
WMOs -	12 x 5400W =	64,800W
WMOs -	10 x 5800W =	58,000W
	76	489,500W (489.5kW)

Per Column B

Based upon 76 cooking appliances, the demand factor per Column B is 16 percent (.16) and when applied to the total kW rating yields the following demand load:

489.5kW x .16 = 78.32kW (78,320W)

Per Column C - Determine the maximum demand of appliances and compare with the calculated value per Column B. Use the lowest value.

Max Demand for 76 Appliances = 25kW + (.75kW x 76) = 82kW (82,000W)

Applying the lowest value, the service demand for the cooking appliances is 78.32kW.

Now, if the cooking appliances were all ranges, NEC 220.55 could be applied. Observe.

1. Divide the number of ranges by the number of phases.

$\dfrac{15 \text{ ranges}}{3} = 5$ $\qquad\qquad$ $\dfrac{12 \text{ ranges}}{3} = 4$ (consider for 6.3kW and 5.4kW ranges)

$\dfrac{17 \text{ ranges}}{3} = 5.67$ (rounded up to nearest whole number) $= 6$

$\dfrac{10 \text{ ranges}}{3} = 3.33$ (rounded up to nearest whole number) $= 4$ (consider for 6.7kW and 5.8kW ranges)

2. Determine the maximum number of ranges connected between any two phases.

5 ranges x 2 = 10 ranges $\qquad$ 4 ranges x 2 = 8 ranges $\qquad$ 6 ranges x 2 = 12 ranges
     (6.1kW) $\qquad\qquad$ (6.3kW, 5.4kW, 6.7kW, 5.8kW) $\qquad$ (7.8kW)

3. Determine the demand load of ranges per demand factor in Column B. The appliances total kW rating (connected load) must first be determined.

```
Ranges -   6.1kW x 10  =   61.0kW
Ranges -   6.3kW x  8  =   50.4kW
Ranges -   5.4kW x  8  =   43.2kW
Ranges -   7.8kW x 12  =   93.6kW
Ranges -   6.7kW x  8  =   53.6kW
Ranges -   5.8kW x  8  =   46.4kW
                   54      348.2kW (348,200W) [total kW rating]
```

Per Column B

54 Ranges = .18 Demand Factor
348.2kW x .18 = 62.676kW (62,676W)

Determine the maximum demand of ranges per Column C and compare with that calculated per Column B. Use the lower calculated value.

Max Demand for 54 Ranges = 25kW + (.75kW x 54) = 65.5kW (65,500W)

The demand load as determined per Column B (62.676kW) must be used as the service demand load per two phases opposed to the maximum demand per Column C (65.5kW).

4. Determine the single-phase (1φ) and three-phase (3φ) service demand loads.

$\dfrac{62.676 \text{kW}}{2}$ = **31.34kW** [service demand load per phase **(1φ)**]

31.34kW x 3 = **94.02kW** [three-phase **(3φ)** service demand load]

220.56 and Table - Kitchen Equipment - Other Than Dwelling Unit(s) (71. - 72.) [2]

71. Determine the demand load of the following commercial appliances that will be used in a 24-hour restaurant currently under construction.

3 - 15kW cooktop grills	4 - 10kW ovens	3 - 5kW deep fryers
2 - 3.75kW dishwasher booster heaters	2 - 11kW water heaters	3 - 3kW coffee makers
	2 - 2.4kW toasters	

When calculating the demand load of commercial cooking equipment, NEC and Table 220.56 must be referenced. NEC 220.56 permits the use of the demand factors listed in Table 220.56 when computing the load for electric cooking equipment, dishwasher booster heaters, water heaters and other kitchen equipment that has thermostatic control or *intermittent use. The demand factors of Table 220.56 do not apply to space heating, ventilating or air-conditioning equipment.

Using the data provided for the cooking equipment, the demand load can be computed based on the number of units of equipment and the equipment's total connected load.

Cooktop grills	3 x 15KW	= 45.0kW
Ovens	4 x 10kW	= 40.0kW
Deep fryers	3 x 5kW	= 15.0kW
Booster Heaters	2 x 3.75kW	= 7.5kW
Water heaters	2 x 11kW	= 22.0kW
Coffee makers	3 x 3kW	= 9.0kW
Toasters	2 x 2.4kW	= 4.8kW
19 (units)		143.3kW (connected load)

The demand factor for 6 or more units of equipment per Table 220.56 is 65 percent (.65). Therefore, the cooking equipment's demand load is,

$$143.3kW \times .65 = 93.15kW$$

NEC 220.56 also states that the feeder or service demand load must not be less than the sum of the largest two kitchen equipment loads. The two largest units of kitchen equipment are two of the cooktop\grills which totals 30kW. In comparison, the cooking equipment's demand load remains at 93.15kW.

* As applied to kitchen equipment, the word *intermittent* does not carry the same meaning as referenced in Article 100 as it pertains to duty cycle. Per NEC 220.56, the wording *intermittent use* refers to equipment that performs an "irregular use" or "not always in use".

72. In the kitchen of a small convenient store which has take-out services the following 240V, single-phase cooking equipment is being fed from a sub-panelboard:

1 - 2.5kW deep fryer	1 - 4kW oven	1 - 3.3kW grill	1 - 1.2kW coffee maker
	1 - 11.5kW water heater	1 - 12kW dishwasher booster heater	

Determine the feeder demand load and the size 75°C copper feeder conductors needed to supply the equipment. All affiliated equipment is rated for 75°C.

The total connected load of the cooking equipment is,

$$2.5kW + 4kW + 3.3kW + 1.2kW + 11.5kW + 12kW = 34.5kW$$

Applying the demand factor of Table 220.56 for 6 units of equipment,

$$34.5kW \times .65 = 22.43kW$$

Considering the largest two kitchen equipment loads,

$$11.5kW \text{ (water heater)} + 12kW \text{ (booster heater)} = 23.5kW$$

Because the two largest kitchen equipment loads exceeds the calculated demand load the feeder demand load is based on 23.5kW. The feeder demand load in amperes is,

$$23.5kW / 240V = 97.92A$$

Referring to Table 310.15(B)(16), as a minimum, 3 AWG conductors which have a rated ampacity of 100 amps at 75°C is required to be used as the feeder conductors.

220.60 - Noncoincident Loads (73. - 74.) [2]

73. The nameplate rating of a household dishwasher rated for 120V list the following loads:

Motor - 1.8A   Heater - 6.7A

Determine the dishwasher load that would contribute to calculating a service load.

The motor and heater loads listed on the nameplate of the dishwasher reflect the use of two separate cycles. The motor load is used exclusively for the washing and rinsing cycle of the dishwasher while the heater load is used exclusively for the drying cycle. Being that both cycles will never operate simultaneously (at the same), they are considered noncoincident (not occurring at the same time) loads according to NEC 220.60. When computing the feeder or service load, NEC 220.60 only requires the largest load that will operate at one time to be used where in this case is the dishwasher's heater load. Therefore, the dishwasher load that will contribute to the service load is 6.7A or 804VA (6.7A x 120V).

74. The air-conditioning and heating loads in a commercial building operates from a split-system. The air-conditioning load totals 43.5kVA while the heating load totals 55kW. Are both loads required to be totaled to determine the building's service demand load?

Because both air-conditioning and heating loads will never operate simultaneously, only the larger load of the two is required to be used in determining the building's service demand load according to NEC 220.60. For this building, the 55kW heating load must be used in calculating the service demand load.

220.61 - Feeder or Service Neutral Load

220.61(A), *Exception* - (Basic Calculation)

75. The maximum unbalanced load between one ungrounded conductors on a 5-wire, 2-phase system is 2377W. Determine the demand load of the neutral conductor.

The *Exception* to NEC 220.61(A) requires the demand load of a neutral conductor that originates from a 3 or 5 wire, 2-phase system to be increased by 140 percent. As a result, the neutral conductor will realize a 3327.8W (2377W x 1.40) demand load.

220.61(B)(1) - (Permitted Reductions) (76. - 77.) [2]

76. In a single-family dwelling an electric dryer and range are used instead of gas appliances. The dryer is rated for 6500W and the range is rated for 11.8kW. Determine the neutral load of both appliances. Determine both appliances contribution to this dwellings service neutral load.

According to NEC 220.61(B)(1), when household electric ranges, wall-mounted ovens, counter-mounted cooking units and electric dryers are supplied from ungrounded feeder or service conductors, the neutral conductor (which carries the unbalanced load between the ungrounded conductors) can be calculated at 70 percent of the load on the ungrounded (neutral) conductors. Tables 220.54 (dryers) and 220.55 (household cooking appliances) can be used in determining the service (or feeder) demands of the given appliances.

NEC 220.54 requires the demand load of a household electric dryer to be 5000W or larger. Table 220.54 requires the demand load of one dryer to be rated at 100 percent which in this case is 6500W. Therefore, the service neutral load contributed by the dryer is,

$$6500W \times .70 = 4550W$$

Since the range is rated for 11.8kW, the demand load is taken from Column C of Table 220.55 which is 8kW (8000W). Therefore, the service neutral load contributed by the range is,

$$8000W \times .70 = 5600W$$

77. A 45 unit apartment complex is furnished with 4.5kW dryers (240/120V), 7.7kW cooktops (240V) and 6.8kW wall-mounted ovens (wmo) (240/120V). Determine the service neutral load for these appliances.

Again, NEC 220.54 requires the load of a household electric dryer to be 5000W or larger. Using a load of 5000W per dryer and the demand factor (.25) listed in Table 220.54, the service load for the dryers is calculated at 56,250W (5000W x 45 x .25) which leads to the service neutral load for the dryers,

$$56,250W \times .70 = \textbf{39,375W (39.375kW)}$$

Table 220.55, **Note 4** permits a cooktop and one wall-mounted oven to be treated as one range, this provision does not apply because the question involves determining the service neutral demand load for a 45 unit complex opposed to a single complex where a cooktop/wmo combination is required to be installed in the same room. Now since the cooktops are rated solely for 240V no neutral voltage is involved. Being the case, only the 45 wall-mounted ovens are considered to determine the service neutral load but first the service load must be determined.

Per **Column B** to Table 220.55 the service load is,

$$6.8kW \text{ (wall-mounted oven)} \times 45 \times .20 = 61.2kW$$

Per **Column C** to Table 220.55 the service (maximum demand) load is,

$$45\text{-}6.8kW \text{ wall-mounted ovens} = 25kW + (.75kW \times 45) = 58.75kW$$

Using the lowest service load (58.75kW) of the two calculated values the service neutral load for the cooking appliances is,

$$58.75kW \times .70 = \textbf{41.125kW}$$

Based on each service neutral load per household appliances, the total service neutral load is **80.5kW** (39.375kW + 41.125kW).

220.61(B)(2) - (Permitted Reductions)

78. What size 75°C copper conductors are required to serve as the service neutral for the appliances in question No. 77. if the incoming service is rated for 240/120V?

The service neutral load in amperes calculated is,

$$80,500W / 240V = 335.42A$$

NEC 220.61(B)(2) permits the neutral load of a 3-wire single-phase *ac* system (240/120V) to be reduced to 70 percent for that portion of the neutral (unbalanced) load exceeding 200 amps. This means that the calculated neutral load (335.42A) is allowed to be decreased before sizing the service neutral conductors. As a result, the service neutral load is so calculated,

$$\begin{array}{r} 335.42A \\ \underline{-200.00A} \\ 135.42A \text{ (neutral load exceeding 200 amps)} \end{array}$$

$$200A + 94.79A \ (135.42A \times .70) = 294.79A \text{ (service neutral load)}$$

Based on the reduced service neutral load, 350 kcmil copper conductors rated for 75°C [Table 310.15(B)(16)] at 310 amps are required to serve as the service neutral conductors.

220.61(C)(1) - (Prohibited Reductions)

79. What size 60°C copper conductors are required to serve as the neutral conductors for the electric dryer and range in question No. 76. if the appliances are rated for 208/120V-1ϕ and located in an apartment complex?

Although it may seem as though this application would reference the use of NEC 220.61(C)(1), [3-wire circuit consisting of 2 ungrounded conductors and the neutral conductor of a 4-wire, 3-phase, wye-connected system] it does not. This provision pertains to feeder or service neutral loads and not branch circuit loads such as these appliances would require. Therefore, the 70 percent reduction of each appliance branch-circuit neutral conductor is permitted per NEC 220.61(B)(1). As a result, each appliance neutral load is calculated based on the results found in question No. 76.

$$\text{(Dryer) (4550W) / 208V} = 21.88A$$
$$\text{(Range) (5600W) / 208V} = 26.92A$$

According to Table 310.15(B)(16) at 60°C, a 10 AWG copper conductor is required to serve as the dryer and range neutral conductors. NEC 210.19(A)(3) - *Exception* 2, reference the permitted use of the neutral conductor of a 3-wire branch circuit supplying a household electric range, a wall-mounted oven, or a counter-mounted cooking unit.

220.61(C)(1) and **(2)** - (Prohibited Reductions)

80. A 208/120V, 3-phase, 4-wire service is required to supply the following loads:

   Three-phase motors (208V) - 193.5A per phase
   Cooking Equipment (208/120V, 1ϕ) - $L_1 = 81A$, $L_2 = 83A$
   Incandescent lights (120V) - $L_1 = 115A$, $L_2 = 123A$, $L_3 = 117A$
   Fluorescent lights (120V) - $L_3 = 229A$
   Receptacles (120V) - $L_2 = 108A$, $L_3 = 113A$

Determine the service neutral demand load.

According to NEC 220.61(C)(2), when a neutral load consist of nonlinear loads that are supplied from a 4-wire, wye connected, 3-phase system, that portion of the neutral load which consist of nonlinear loads cannot be reduced by 70 percent.

In this problem, the fluorescent lights are considered nonlinear loads. Nonlinear loads produce a third harmonic frequency which causes additive phase currents to flow through the neutral conductor of a 208/120V or 480/277V, 4-wire, wye connected, 3-phase system unlike a 240/120V single-phase system or a 240/120V, 4-wire, three-phase system where only the unbalanced current between the ungrounded conductors flows through the neutral conductor.

Because these additive phase currents produce an excessive amount of current flow through the neutral conductor, the conductor is quite often damaged due to excessive overheating if not properly sized. For more detailed information on nonlinear loads, harmonic current and similar problems refer to NEC 310.15(B)(5) - Neutral Conductor (Volume 2).

Although the cooking equipment is not considered a nonlinear load, a 70 percent reduction of the equipment's grounded (neutral) conductor is not permitted also according to NEC 220.61(C)(1), because the equipment would be supplied by a 3-wire circuit consisting of two phase wires (208/120V, 1ϕ) and the neutral of a 4-wire, 3-phase, wye connected system.

Now to determine the service neutral demand load all neutral loads are considered. Because the 3-phase motors are 3-wire, line-to-line loads they will not contribute to the neutral loads.

Cooking Equipment

The cooking equipment's neutral current is computed using the following formula (to approximate the cooking equipment's neutral current),

$$I_N = \sqrt{(L_1^2 + L_2^2 + L_3^2) - (L_1 \times L_2 + L_2 \times L_3 + L_1 \times L_3)}$$

$$I_N = \sqrt{(81^2 + 83^2 + 0^2) - (81 \times 83 + 83 \times 0 + 81 \times 0)}$$

$$= \sqrt{(6{,}561 + 6{,}889) - (6{,}723)}$$

$$= \sqrt{(13{,}450) - (6{,}723)}$$

$$= \sqrt{6{,}727} = \textbf{82A}$$

As you can see, the calculated neutral current will carry approximately the same amount of current as the ungrounded conductors of this 3-wire circuit and not an unbalanced current. NEC 310.15(B)(5)(b) refers to this conductor as a common conductor opposed to a neutral conductor. If this was referred to as a neutral conductor then the conductor would be expected to carry only 2 amperes, the unbalanced current between the two ungrounded conductors, L_1 and L_2.

Consideration is now given to that portion of the neutral loads that can be derated and those that cannot. Again, the previously given formula is used.

Incandescent lights - 115A (L_1), 123A (L_2), 117A (L_3)
Receptacles -   +   108A (L_2), 113A (L_3)
Total - 115A (L_1), 231A (L_2), 230A (L_3)

$$I_N = \sqrt{(115^2 + 231^2 + 230^2) - (115 \times 231 + 231 \times 230 + 115 \times 230)}$$

$$= \sqrt{(13{,}225 + 53{,}361 + 52{,}900) - (26{,}565 + 53{,}130 + 26{,}450)}$$

$$= \sqrt{(119{,}486) - (106{,}145)}$$

$$= \sqrt{13{,}341} = \textbf{116A (rounded up)}$$

As you can see the neutral current per incandescent lights and receptacles loads totals less than 200 amps therefore, the 70 percent reduction cannot be applied. The approximate neutral current for both loads is **116A**.

Other neutral loads

$$
\begin{array}{lr}
\text{Fluorescent lights -} & 229\text{A} \\
\text{Cooking Equipment -} & \underline{82\text{A}} \\
& \mathbf{311A}
\end{array}
$$

The service neutral demand load is determined by totaling all neutral currents.

$$116A + 311A = \mathbf{427A}$$

ARTICLE 220 - STANDARD LOAD CALCULATIONS

The following procedures for calculating service loads for residential, commercial and industrial buildings are based on applicable sections of **Part III** of Article 220 as referenced.

Standard Load Calculations for One-Family Dwellings (81.- 85.) [5]

Refer to **(Worksheet A** - Volume **4)** STANDARD LOAD CALCULATIONS FOR ONE-FAMILY DWELLING for related questions.

81. The computed floor area of a single-family house is 2435 square feet. The house has an 800 square feet unfinished basement that's adaptable for future use. If this house will utilize 5 small-appliance circuits and one laundry circuit, determine the general lighting, receptacle, small-appliance and laundry demand load based on the information provided.

1. General Lighting and Receptacle Loads <NEC 220.12, Table 220.12, 220.14(J) and 220.42> (Open porches, garages, unused or unfinished spaces not adaptable for future use not included.)

Because the unfinished basement is adaptable for future use it must be included with the total square footage of the house.

$$(2435SF + 800SF) \times 3VA = 9705VA$$

2. Small-Appliance Circuit Load (Portable) <NEC 220.52(A)>

$$1500VA \times 5 = 7500VA$$

3. Laundry Circuit Load <NEC 220.52(B)>

$$1500VA \times 1 = \underline{1500VA}$$

TOTAL (Lines 1. – 3.) = 18,705VA
(If Total VA is less than or equal to 120,000VA, step c. is not required)

APPLY DEMAND FACTORS <NEC 220.42 and Table 220.42>

a. First 3000VA of above TOTAL (At 100%) = 3000.00VA

b. 15,705VA   x .35   = 5496.75VA
(Total VA – 3001VA up to 117,000VA)

c.   =   <u>0</u>

TOTAL (Lines a. - c.)   = 8496.75VA

The general lighting, receptacle, small-appliance and laundry demand load for this house is 8496.75VA.

82.   A 4200 SF dwelling unit being supplied by a 230/115V single-phase source has the following loads:

230V	**115V**
10kW cooktop	1kW dishwasher
5kW water heater	1600W microwave oven
15kW electric heat (2)	1500W HVL (4)
AC - 29A compressor, 1.5A fan motor	960VA disposal
AC - 23.5A compressor, 1.5A fan motor	small-appliance circuits (6)
745VA blower motor (2)	

230/115V
8.5kW wall-mounted ovens (WMO) (2)
4.5kW dryer

Use the standard load calculation to determine the service and neutral loads.

1. General Lighting and Receptacle Loads <NEC 220.12, Table 220.12, 220.14(J) and 220.42> (Open porches, garages, unused or unfinished spaces not adaptable for future use not included.)

$$4200 \text{ SF x 3VA } = \ 12{,}600\text{VA}$$

2. Small-Appliance Circuit Load (Portable) <NEC 220.52(A)>

$$1500\text{VA x 6} = \quad 9000\text{VA}$$

3. Laundry Circuit Load <NEC 220.52(B)>

$$1500\text{VA x 1} = \quad \underline{1500\text{VA}}$$

TOTAL (Lines 1. – 3.) = 23,100VA
(If Total VA is less than or equal to 120,000VA, step c. is not required)

APPLY DEMAND FACTORS <NEC and Table 220.42>

a. First 3000VA of above TOTAL (At 100%) =   3000VA

b. 20,100 VA   x .35   =   7035VA
    (Total VA - 3001 VA up to 117,000VA)

c.   =   ___0___

TOTAL (Lines a. - c.)   = 10,035VA

GENERAL LIGHTING and RECEPTACLE, SMALL-APPLIANCE and LAUNDRY LOADS

1. - 3. <u>LINE LOAD</u>   <u>NEUTRAL LOAD</u>

     10,035VA   10,035VA

4. Appliance Loads (Fastened-In-Place) <NEC 220.53> (Use nameplate rating of each appliance. Electric ranges, dryers, space-heating equipment or air-conditioning equipment not included.)

115V Appliances	**VA Rating**
1. Dishwasher	1000
2. Microwave	1600
3. HVL x 4 (@ 1500W)	6000
4. Disposal	<u>960</u>
	9560 (Total 115V Appliances)

230V Appliances	**VA Rating**
1. Water Heater	<u>5000</u> (Total 230V Appliance)

     APPLIANCES TOTAL   = 14,560VA

APPLY DEMAND FACTOR (Applicable, when number of above appliances exceeds four (4) or more.)

$$\text{(Appliances Total) } \underline{14,560VA} \times .75 = 10,920VA$$

APPLIANCE LOADS

4.   <u>LINE LOAD</u>   <u>NEUTRAL LOAD</u> (Refer to condition)

     10,920VA   7170VA

<u>Condition</u>
Because the water heater is a 230 volts line-to-line load it will not contribute to the Neutral load as the other appliances and therefore must not be included. As a result, the NEUTRAL LOAD is equal to,

$$14,560VA - 5000VA \times .75 = \underline{7170VA}$$

The demand factor could still be applied because more than four appliances were remaining after the exclusion of the water heater.

5. Clothes Dryer <NEC and Table 220.54> (Use 5000W [VA] or nameplate rating, whichever is larger.)

CLOTHES DRYER

5.	LINE LOAD	NEUTRAL LOAD [NEC 220.61(B)(1)]
	5000VA	3500VA (5000VA x .70 of line load)

6. Electric Ranges and Other Cooking Appliances <NEC and Table 220.55>

Minimum Demand Load determined per Column C and Table 220.55, **Notes 4** and **1**.

Applying Column C

          3 appliances under 12kW – Maximum demand = 14kW

Applying **Note 4** (treated as one range)

Cooktop -	10,000W
WMO (2) @ 8.5kW each -	17,000W
	27,000W

Applying **Note 1** (when treated as one range)

1. 27kW – 12kW   = 15kW
2. 15 x .05   = .75 + 1 = 1.75
3. 8kW (Col. C) x 1.75 = 14kW (kVA)

Both Column C and the provisions of **Note 4** (utilizing **Note 1**) produces the same results, therefore the minimum demand load is 14kW (14,000VA).

RANGE/COOKING APPLIANCES

6.	LINE LOAD	NEUTRAL LOAD [NEC 220.61(B)(1)]
	14,000VA	7700VA

Because the cooktop is a 230 volts line-to-line load it will not contribute to the neutral load as the other appliances and therefore must not be included. As a result, the range/cooking appliances LINE LOAD must be recalculated omitting the cooktop to determine the NEUTRAL LOAD.

For 2 appliances – WMOs
    Per Column C – Maximum demand = 11kW
    Per Column B – 8.5kW x 2 x .65   = 11.05kW

Using the lowest calculated demand load as the line load the NEUTRAL LOAD is,

$$11kW \times .70 = 7.7kW \ (7700VA)$$

7. Heating and Air-Conditioning (AC) Equipment <NEC References - 220.50, 220.51, 220.60, Table 430. 248 and 440.6(A)> (Include VA rating of air handler [blower].)

Heating Load

    Electric Heat
    15kW x 2 = 30kW (30,000VA) (Total Heat)

AC Load

    Compressor(s)   Fan Motor(s)
    230V x 29A + 230V x 1.5A   =   7015VA
    230V x 23.5A + 230V x 1.5A   =   5750VA
                  Total AC = 12,765VA

Heating load is larger.

           Electric Heat   Blower
Line Load = 30,000VA + 745VA x 2 = 31,490VA

HEATING and AC

7.	LINE LOAD	NEUTRAL LOAD
	31,490VA	0

8. Largest Motor <NEC 430.17, 430.24, 440.7 and 440.33> (Use motor with highest full-load current (FLC) regardless of voltage rating.)

    Largest Motor (LM) = 230V x 29A = 6670VA - AC compressor

6670VA x .25 = 1668VA (rounded-up)
(LM)

LARGEST MOTOR

8.	LINE LOAD	NEUTRAL LOAD (115V motors only)
	1668VA	0

TOTAL DEMAND LOAD (LINE and NEUTRAL) (Add lines 1. - 8.)

	LINE LOAD	NEUTRAL LOAD
1. - 3.	10,035VA	10,035VA
4.	10,920VA	7170VA
5.	5000VA	3500VA
6.	14,000VA	7700VA
7.	31,490VA	0
8.	1668VA	0
	73,113VA	28,405VA

DWELLING'S OPERATING LINE VOLTAGE - 230V
(Given operating voltage or as determined per test examination)

CALCULATE MINIMUM LINE and **NEUTRAL LOADS**
(Divide Total Demand Load **[VA]** by operating line voltage **[V]**)

LINE LOAD   = 73,113VA / 230V = 317.88A
NEUTRAL LOAD = 28,405VA / 230V = 123.5A

83. A 9700SF dwelling unit being supplied by a 240/120V single-phase source has the following loads:

240V
6.5kW cooktop
6.8kW cooktop
AC - 25.3A compressor, 1.2A fan motors (2)
AC - 20.2A compressor, 1.2A fan motors (2)
AC - 14.4A compressor, 1.1A fan motor
HT - 20kW, 14.8A blower (2)
HT - 15kW, 10.3A blower (2)
HT - 10kW, 10.3A blower
4.5kW water heater (3)
Swimming pool motor
    cleaning - ¾ HP
    circulatory - 1½ HP

240/120V
6.7kW dryers (2)
5.5kW wall-mounted ovens (2)
6.4kW wall-mounted oven
2.2kW deep fryer
Hot Tub Spa
    5.5kW heater (240V)
    1½ HP circulatory motor (120V)

120V
1kW dishwasher (2)
2000W rotisserie
1600W microwave oven (2)
1250W microwave oven
13.3A upright freezer
1500W heat-vent-lights (6)
960VA disposal
780VA disposal
756VA trash compactors (2)
½ HP gate opener
¾ HP sump pump motor
1 HP central vacuum motor
1½ HP fountain pump

small-appliance circuits (8)
laundry circuits (2)

Use the standard load calculation to size the service and neutral loads, and the service, neutral and grounding electrode conductors where copper conductors at 75°C are used.

1. General Lighting and Receptacle Loads <NEC 220.12, Table 220.12, 220.14(J) and 220.42> (Open porches, garages, unused or unfinished spaces not adaptable for future use not included.)

$$9700SF \times 3VA = \quad 29,100VA$$

2. Small-Appliance Circuit Load (Portable) <NEC 220.52(A)>

$$1500VA \times 8 = \quad 12,000VA$$

3. Laundry Circuit Load <NEC 220.52(B)>

$$1500VA \times 2 = \quad \underline{3000VA}$$

TOTAL (Lines 1. – 3.) = 44,100VA
(If Total VA is less than or equal to 120,000VA, step c. is not required)

APPLY DEMAND FACTORS (NEC and Table 220.42)

a. First 3000VA of above TOTAL (At 100%) = 3000VA

b. 41,100 VA   x .35   = 14,385VA
   (Total VA – 3001 VA up to 117,000VA)

c.   = __0__

      TOTAL (Lines a. - c.)   = 17,385VA

GENERAL LIGHTING and RECEPTACLE, SMALL-APPLIANCE and LAUNDRY LOADS

1. - 3. <u>LINE LOAD</u>   <u>NEUTRAL LOAD</u>

    17,385VA   17,385VA

4. Appliance Loads (Fastened-In-Place) <NEC 220.53> (Use nameplate rating of each appliance. Electric ranges, dryers, space-heating equipment or air-conditioning equipment not included.)

120V Appliances	**VA Rating**
1. Dishwashers (2)	2000
2. Rotisserie	2000
3. Microwaves (3)	4450
4. Freezer (120V x 13.3A)	1596
5. Heat-Vent-Lights (6)	9000

6. Disposals (2)	1740
7. Trash Compactors (2)	1512
8. Gate Opener (120V x 9.8A*)	1176
9. Sump Pump (120V x 13.8A*)	1656
10. Vacuum mtr. (120V x 16A*)	1920
11. Fountain Pump (120V x 20A*)	2400
12. Circ. mtr. (Spa) (120V x 20A*)	2400
	31,850 (Total 120V Appliances)

240V Appliances	**VA Rating**
1. Water Heaters (3)	13,500
2. Heater (Spa)	5500
3. Cleaning mtr. (Pool) (240V x 6.9A*)	1656
4. Circ. mtr. (Pool) (240V x 10A*)	2400
5. Deep fryer	2200
* - Table 430.248	25,256 (Total 240V Appliances)

APPLIANCES TOTAL   =   57,106VA

APPLY DEMAND FACTOR (if applicable) - (When number of above appliances exceeds four (4) or more.)

(Appliances Total) 57,106VA x .75 =   42,829.5VA

APPLIANCE LOADS

4.   LINE LOAD   NEUTRAL LOAD (Refer to condition)

42,829.5VA   29,662.5VA

Condition

Because the water heaters and the pool motors are 240 volts line-to-line loads they will not contribute to the neutral load as the other appliances and therefore must not be included.

As a result, the NEUTRAL LOAD is equal to,

57,106VA – 17,556VA x .75 = 29,662.5VA

5. Clothes Dryer <NEC and Table 220.54> (Use 5000W [VA] or nameplate rating, whichever is larger.)

CLOTHES DRYER (2)

5.   LINE LOAD   NEUTRAL LOAD [NEC 220.61(B)(1)]

13,400VA   9380VA (13,400VA x .70 of line load)

6. Electric Ranges and Other Cooking Appliances <NEC and Table 220.55>

Applying Column B

(5.5kW x 2 + 6.4kW + 6.5kW + 6.8kW) x .45 = 13.82kW (rounded-up)

Applying Column C

5 appliances under 12kW – Maximum demand = 20kW

Per Column B, 13.82kW is the lower demand load compared to Column C

RANGE/COOKING APPLIANCES

6.   LINE LOAD   NEUTRAL LOAD NEC 220.61(B)(1)]

13,820VA   6700VA

Because the cooktops are 240 volts line-to-line loads they will not contribute to the neutral load as the other appliances and therefore must not be included. As a result, the range/cooking appliances LINE LOAD must be recalculated omitting the cooktops to determine the NEUTRAL LOAD.

For 3 appliances – WMOs

Per Column C – Maximum demand = 14kW

Per Column B – (5.5kW x 2 + 6.4kW) x .55 = 9.57kW

Using the lowest calculated demand load as the line load the NEUTRAL LOAD is,

9.57kW x .70 = 6.7kW (6700VA) (rounded-up)

7. Heating and Air-Conditioning (AC) Equipment <NEC References - 220.50, 220.51, 220.60, Table 430. 248 and 440.6(A)> (Include VA rating of air handler [blower].)

Heating Load

    Electric Heat
    (1) 20kW (20000VA) x 2 = 40,000VA
    (2) 15kW (15000VA) x 2 = 30,000VA
    (3) 10kW (10000VA)   = 10,000VA
              Total Heat = 80,000VA

AC Load

    Compressors   Fan Motors
    (1) (240V x 25.3A + 240V x 1.2A) x 2 = 12,720VA
    (2) (240V x 20.2A + 240V x 1.2A) x 2 = 10,272VA

(3) 240V x 14.4A + 240V x 1.1A   =   <u>3720VA</u>

Total AC = 26,712VA

Heating load is larger

              Electric Heat   Blowers (1)   Blowers (2)   Blower (3)
Line Load - 80,000VA + 240V x 14.8A x 2 + 240V x 10.3A x 2 + 240V x 10.3A = 94,520VA

HEATING and AC

7.	LINE LOAD	NEUTRAL LOAD
	94,520VA	0

8. Largest Motor <NEC 430.17, 430.24, 440.7 and 440.33> (Use motor with highest full-load current [FLC] regardless of voltage rating.)

Although the circulatory pool motor, the circulatory spa motor and the fountain pump motor are all rated for 1.5 horsepower the largest motor is determined based on the motor with the highest FLC. While the circulatory spa motor and the fountain pump motor have the same FLCs only one consideration is required.

Largest Motor (LM) = 2400VA (20A) @ 120 volts

<u>2400VA</u> x .25 = 600VA
(LM)

LARGEST MOTOR

8.	LINE LOAD	NEUTRAL LOAD (120V motors only)
	600VA	600VA

TOTAL DEMAND LOAD (LINE and NEUTRAL) (Add lines 1. - 8.)

	LINE LOAD	NEUTRAL LOAD
1. - 3.	17,385.0VA	17,385.0VA
4.	41,179.5VA	29,662.5VA
5.	13,400.0VA	9380.0VA
6.	13,820.0VA	6700.0VA
7.	94,520.0VA	0
8.	**600.0VA**	**600.0VA**
	180,904.5VA	63,727.5VA

DWELLING'S OPERATING LINE VOLTAGE - <u>240V</u>
(Given operating voltage or as determined per test examination)

CALCULATE MINIMUM LINE and NEUTRAL LOADS
(Divide Total Demand Load [VA] by operating line voltage [V])

LINE LOAD   = 180,904.5VA / 240V = 753.77A
NEUTRAL LOAD = 63,727.5VA / 240V = ~~265.53~~A* 245.87A

    *Where the neutral load exceeds 200A, NEC 220.61(B)(2) permits the load to be reduced by 70 percent. Complete the following to determine permitted Neutral Demand Load.

    (1) 265.53A – 200A = 65.53A   x .70 = 45.87A
        (Neutral Load)*   (Remainder)   (Permitted Reduction)

    (2) 45.87   A + 200A = 245.87A
        (Permitted Reduction)   (Permitted Neutral Load)

SIZE SERVICE (Size of service based on the calculated LINE LOAD)

SIZE SERVICE REQUIRED (minimum) 800A

SIZING FEEDER/SERVICE CONDUCTORS - (Based on the calculated LINE LOAD and NEC References.) *For* 240/120V, 3-Wire, Single-Phase Dwelling Services and Feeders *up to 400 amperes* only. <NEC References - 310.15(B)(7) and Table 310.15(B)(7)> *For* other [3-Wire, Single-Phase] or [Three-Phase] Services and Feeders. <NEC References - 215.2(A), 230.42(A), 240.4(B) & (C), 310.10(H), 310.15(B)(2) & (3) and Table 310.15(B)(16)>

Because there are no single conductors that can supply the 800A service the use of parallel conductors are required per NEC 310.10(H). The parallel conductors can be run in sets of two or other desired combinations, however when installing parallel conductors in the same raceway just remember that the ampacity of the conductors has to be derated according to NEC 310.15(B)(3)(a) and where applicable NEC 310.15(B)(3)(c) and the correction factors of Table 310.15(B)(2)(a).

In this situation let's use a combination of 3 parallel conductors per line without having to consider NEC 310.15(B)(3)(a) by installing the parallel conductors in three individual raceways. Applying the service rating, the ampacity of each individual parallel conductor can be determined. Therefore,

$$\frac{800A}{3} = 266.67A \text{ (needed ampacity per Line conductor)}$$

Since 75°C copper conductors are required, three 250 kcmil conductors (255A) can be used for the Line (Service) conductors. Although the total ampacity of three 250 kcmil conductors (765A) is less than the 800A service and the needed ampacity per line conductor they will suffice, the use of these conductors are permitted per NEC 240.4(B).

FEEDER/SERVICE CONDUCTORS 3-250 kcmil copper

SIZING NEUTRAL CONDUCTOR - (Based on the calculated NEUTRAL LOAD and NEC References.) <NEC References - 215.2(A)(2), 220.61, 230.42(C), 250.24(C), 310.10(H), 310.15(B)(2), (3) & (5), 310.15(B)(7) and Table 310.15(B)(16)>

Based on the calculated neutral load and the use of parallel service conductors, the neutral conductors must also be installed in parallel per NEC 310.10(H)(1). Because the service conductors will consist of three parallel conductors per phase, the neutral conductors must also consist of three parallel conductors based on the calculated neutral load being divided by three which results to 81.96A (245.87A / 3). Per Table 310.15(B)(16), at 75°C, a 4 AWG copper conductor which has a rated ampacity of 85A will satisfy the calculated neutral load being distributed amongst three such conductors. However, because NEC 310.10H(1) only allows a 1/0 AWG conductor or larger to be installed in parallel, three 1/0 AWG copper conductors must be used instead as the parallel neutral conductors.

NEUTRAL CONDUCTOR(S) 3-1/0 AWG copper conductors

See question Nos. 1. - 8. of Article 250 (Volume 2) for further reference when the grounded (neutral) conductor is not used as a circuit conductor.

SIZING GROUNDING ELECTRODE CONDUCTOR <NEC References - 250.24(C), 250.66 & Table 250.66 and Table 8 of Chapter 9>

Based on the total cross-sectional area of the selected copper service conductors (750 kcmil copper), per Table 250.66, a 2/0 AWG copper grounding electrode conductor must be used.

Although NEC 250.24(C)(1) requires the neutral (grounded) conductor to be either the same size or larger than the grounding electrode conductors, the total circular mils of the three 1/0 AWG neutral conductors (316,800 cmils [105,600 cmils x 3]) exceeds the circular mils of a 2/0 AWG conductor (133,100 cmils) per Table 8 of Chapter 9. As a result, this classifies the use of the three 1/0 AWG parallel neutral conductors as being larger than the 2/0 AWG grounding electrode conductor.

Remember only one single conductor is required to serve as the grounding electrode conductor opposed to three.

GROUNDING ELECTRODE CONDUCTOR 1-2/0 AWG copper

In summary, the line and neutral loads for this single-family dwelling per standard load calculation is 753.77A and 245.87A respectively. As a minimum, an 800A service is required. Three (3) 250 kcmil copper conductors were selected to serve as the service conductors, three (3) 1/0 AWG copper conductors were selected to serve as the neutral conductors and one (1) 2/0 AWG copper will be used to serve as the grounding electrode conductor.

84. Re-calculate the Heating and AC load in question No. 83. if gas furnaces were used instead of electric furnaces. The gas furnaces are equipped with the following blowers and are rated for 120V:

2 - 10.8A Blowers   2 - 9.6A Blowers   1 - 7.7A Blower

Heating Load (Gas)

AC Load

<pre>
        Compressors   Fan Motors
(1) 240V x 25.3A + 240V x 1.2A x 2 = 12,720VA
(2) 240V x 20.2A + 240V x 1.2A x 2 = 10,272VA
(3) 240V x 14.4A + 240V x 1.1A   = 3,720VA
                                      26,712VA
</pre>

AC load is largest.

<pre>
            AC   Blowers (1)   Blowers (2)   Blower (3)
Line Load = 26,712VA + 120V x 10.8A x 2 + 120V x 9.6A x 2 + 120V x 7.7A = 32,532VA
</pre>

When re-calculated, the AC load will undoubtedly become the largest load between the heating and AC loads. The only significant load the gas furnaces will contribute towards the electrical load are by means of the 120V blowers. The AC load will contribute 32,532VA towards the line load and 5820VA (total VA of 120V blowers) towards the neutral load.

85. Recalculate the Heating and AC load in question No. 83. if the given heating and air-conditioning equipment was used as a heat pump system.

Where a heat pump system is used the heating and air-conditioning (AC) loads must be combined because both loads could eventually run simultaneously (at the same time). Therefore, the total Heating and AC load as applied to question No. 83. is as follows,

Heating Load = 80,000VA
AC Load   = 26,712VA

<pre>
             Heat   AC   Blowers (1)   Blowers (2)
Line Load = 80,000VA + 26,712VA + 240V x 14.8A x 2 + 240V x 10.3A x 2
               Blower (3)
           + 240V x 10.3A = 121,232VA
</pre>

Using a heat pump system the line load is increased to 121,232VA and the neutral load still remains at zero (0) because none of the heating and AC equipment is rated for 120V.

Standard Load Calculations for Multifamily Dwellings (86. - 88.) [3]

Refer to **(Worksheet C** - Volume 4) STANDARD LOAD CALCULATIONS FOR MULTIFAMILY DWELLING for related questions.

86. A 28 unit apartment complex (925-SF of living space) is supplied from a 240V/120V single-phase service and contains the following (electric) loads:

240V	120V
10kW heating unit	1000W microwave oven
5kW cooktop	650VA trash compactor
4500W dryer*	760VA disposal
4700VA AC	1200W dishwasher (1kW ht/1.67A mtr)
4.7kW wall-mounted cooking unit*	1500W Heat-Vent-Light
5kW water heater	
560W blower	
* - 240/120V	

Use the standard load calculation to determine the service and neutral loads.

1. General Lighting and Receptacle Loads <NEC 220.12, Table 220.12, 220.14(J) and 220.42> (Open porches, garages, unused or unfinished spaces not adaptable for future use not included.)

$$925SF \times 28 \times 3VA = \quad 77,700VA$$

2. Small-Appliance Circuit Load (Portable) <NEC 220.52(A)> (Minimum of two circuits required.)

$$1500VA \times 2 \times 28 = \quad 84,000VA$$

3. Laundry Circuit Load <NEC 220.52(B)>

$$1500VA \times 28 = \quad \underline{42,000VA}$$

TOTAL (Lines 1. - 3.) = 203,700VA
(If Total VA is less than or equal to 120,000VA, step c. is not required)

APPLY DEMAND FACTORS <NEC 220.42 and 220.42>

a. First 3000VA of above TOTAL (At 100%) = 3000VA

b. 117,000VA   x .35   = 40,950VA
   (Total VA – 3001VA up to 117,000VA)

c. 83,700VA   x .25   = 20,925VA
  (Remainder of TOTAL exceeding 120,000VA)*

  * (203,700VA – 120,000 VA = 83,700VA)

TOTAL (Lines a. - c.)   = 64,875VA

GENERAL LIGHTING and RECEPTACLE, SMALL-APPLIANCE and LAUNDRY LOADS

1. - 3. <u>LINE LOAD</u>   <u>NEUTRAL LOAD</u>

    64,875VA   64,875VA

4. Appliance Loads (Fastened-In-Place) <NEC 220.53> (Use nameplate rating of each appliance. Electric ranges, dryers, space-heating equipment or air-conditioning equipment not included.)

120V Appliances	VA	No. Appliances	Total VA
1. Microwave	1000	28	28,000
2. Trash Compactor	650	28	18,200
3. Disposal	760	28	21,280
4. HVL	1500	28	42,000
5. Dishwasher	1000**	28	<u>28,000</u>
		(Total 120V Appliances)	137,480

**noncoincident load (NEC 220.60) – motor (mtr) used for wash cycle, heater (ht) for drying cycle, heater larger.

240V Appliances	VA	No. Appliances	Total VA
1. Water Heater	5000	28	<u>140,000</u>
		(Total 240V Appliances)	140,000

APPLIANCES TOTAL   = 277,480VA

APPLY DEMAND FACTOR (Applicable, when number of above appliances exceeds four (4) or more.)

(Appliances Total) <u>277,480VA</u> x .75 = 208,110VA

APPLIANCE LOADS

4.   <u>LINE LOAD</u>   <u>NEUTRAL LOAD (Refer to condition)</u>

    208,110VA   103,110VA

Condition
Because the water heaters are 240 volts line-to-line loads they will not contribute to the neutral load as the other appliances and therefore must not be included. As a result, the NEUTRAL LOAD is equal to

277,480VA – 140,000VA x .75 = <u>103,110VA</u>

5. Clothes Dryer <NEC and Table 220.54> (Use 5000W [VA] or nameplate rating, whichever is larger.)

Refer to Table 220.54 - For 28 dryers
Demand Factor (%) = 35% − [.5% x (28 − 23)] = 32.5% (.325)
Dryers - 5000VA x 28 x .325 = 45,500VA

CLOTHES DRYER

5. LINE LOAD   NEUTRAL LOAD [NEC 220.61(B)(1)]

    45,500VA   31,850VA (45,500VA x .70 of line load)

6. Electric Ranges and Other Cooking Appliances <NEC and Table 220.55>

Range Load determined per Columns B and C to Table 220.55.

Per Column B (5kW Cooktops and 4.7kW WMOs) (Note 3)

        Cooktops - 28 x 5000W = 140,000W
        WMOs -   28 x 4700W = 131,600W
                  56   271,600W

    Demand Factor per Column B - 56 Ranges = .18 Demand Percent

    271,600W x .18 = 48,888W(VA)

Per Column C

    Max Demand for 56 Ranges = 25kW + (56 x .75kW) = 67kW(67,000VA)

    Use the smallest demand between Column B (48,888VA) and Column C (67,000VA)

RANGE/COOKING APPLIANCES

6. LINE LOAD   NEUTRAL LOAD [NEC 220.61(B)(1)]
                                        (Refer to condition)

    48,888VA   22,108.8VA

Condition
Because the cooktops are 240 volts line-to-line loads they will not contribute to the neutral load as cooking appliances and therefore must not be included. As a result, the range/cooking appliances LINE LOAD must be recalculated omitting the cooktops to determine the NEUTRAL LOAD.

Per Column B (4.7kW WMOs)

WMOs - 28 x 4700W = 13,1600W

Demand Factor per Column B

28 Ranges = .24 Demand Percent

131,600W x .24 = 31,584W(VA)

Per Column C

Max Demand for 28 Ranges = 15kW + (28 x 1kW) = 43kW(43,000VA)

Using the lowest calculated demand load as the line load the NEUTRAL LOAD is,

31,584kW x .70 = 22.1088kW (22,108.8VA)

7. Heating and Air-Conditioning (AC) Equipment <NEC References - 220.50, 220.51, 220.60, Table 430.248 and 440.6(A)> (Include VA rating of air handler [blower].)

Heating Load

10kW (10,000VA) x 28 = 280,000VA

AC Load

4700VA x 28 =131,600VA

Heating load is larger.

<pre>
            Heat   Blowers
Line Load = 280,000VA + 560VA x 28 = 295,680VA
</pre>

HEATING and AC

7.	LINE LOAD	NEUTRAL LOAD
	295,680VA	0

8. Largest Motor <NEC 430.17, 430.24, 440.7 and 440.33> (Use motor with highest full-load current [FLC] regardless of voltage rating.)

Largest Motor (LM) = 760VA (6.33A) disposal rated for 120 volts

760VA x .25 =190VA
(LM)

LARGEST MOTOR

8. <u>LINE LOAD</u>   <u>NEUTRAL LOAD (120V motors only)</u>

    190VA   190VA

TOTAL DEMAND LOAD (**LINE** and **NEUTRAL**)(Add lines 1. – 8.)

	LINE LOAD	NEUTRAL LOAD
1. - 3.	64,875VA	64,875.0VA
4.	208,110VA	103,110.0VA
5.	45,500VA	31,850.0VA
6.	48,888VA	22,108.8VA
7.	295,680VA	0
8.	190VA	190.0VA
	663,243VA	222,133.8VA

DWELLING'S OPERATING LINE VOLTAGE - <u>240</u>V
(Given operating voltage or as determined per test examination)

CALCULATE MINIMUM LINE and NEUTRAL LOADS
(Divide Total Demand Load **[VA]** by operating line voltage **[V]**)

LINE LOAD   = <u>663,243</u>VA / <u>240</u>V = <u>2763.51A</u>
NEUTRAL LOAD = <u>222,133.8</u>VA / <u>240</u>V = ~~925.56A~~* 707.89A

> *Where the neutral load exceeds 200A, NEC 220.61(B)(2) permits the load to be reduced by 70 percent.
> Complete the following to determine permitted Neutral Demand Load.

(1) <u>925.56</u>   A – 200A = <u>725.56</u>   A x .70 = <u>507.89</u>   A
    (Neutral Load)*   (Remainder)   (Permitted Reduction)

(2) <u>507.89</u>   A + 200A = <u>707.89</u>   A
    (Permitted Reduction)   (Permitted Neutral Load)

The service load for this multifamily dwelling using the standard load calculation is as calculated for both line and neutral loads.

87.   A 54 unit apartment complex is supplied by a 208/120V, 4W, 3-phase system. Each individual unit is supplied by a 208/120V single-phase service. The complex consists of the following units:

650SF units (15)

240-208/120V
8.1/6.1kW range
240-208V
13.3/10kW furnace
2743/3648VA AC
4.5/3.38kW water heater
785/1044VA blower

120V
900VA dishwasher
600VA disposal
1000W microwave
1300W heat-vent-light

1200SF units (12)

240-208/120V
7.18/5.4kW wall-mounted oven

240/208V
8.4/6.3kW cooktop
20/15kW furnace
4114/5472VA AC
4.5/3.38kW water heater
992/1320VA blower

120V
1250VA dishwasher
900VA disposal
780VA trash compactor
1600W microwave
1600W heat-vent-light

900SF units (17)

240-208/120V
10.4/7.8kW range
240-208V
16/12kW furnace
3429/4560VA AC
4.5/3.38kW water heater
884/1176VA blower

120V
900VA dishwasher
600VA disposal
1200W microwave
1600W heat-vent-light

1400SF units (10)

240-208/120V
7.7/5.8kW wall-mounted oven

240/208V
8.9/6.7kW cooktop
24/18kW furnace
4800/6384VA AC
4.5/3.38kW water heater
1065/1416VA blower

120V
1250VA dishwasher
900VA disposal
780VA trash compactor
1800W microwave
1750W heat-vent-light

The cooking appliance loads were taken from question No. 70. The 650SF units are not equipped with washer/dryer connections. A laundry facility is available to all residents and is equipped with gas water heaters. The following house loads will be supplied from a 208/120V, 4W, 3-phase service:

Office - 740SF
12 - 2 x 4 fluorescent light fixtures (1.65A per ballast - 120V)
 5 - 100W Recessed cans (120V)
10 - Duplex receptacles
 8' - Track lighting (120V)
 4 - Desktop Computers (1.3A - 120V)
 2 - All-in-One Laser Printers (3.8A - 120V)
Copier (10.3/13.7A – 240/208V)
Soft drink Machine (10.4A - 120V)
Water Heater (4.5/3.38kW – 240/208V)
Furnace (13.3/10kW – 240/208V)
AC (2743/3648VA – 240/208V)
Blower - ¾HP blower (240/208V)

Laundromat - 400SF
 5 - Duplex receptacles
 6 - 2 x 4 fluorescent light fixtures (1.65A per ballast - 120V)
12 - Three-phase commercial washers - (13A - 208/120V)
 6 - Three-phase commercial clothes dryers - (208/120V)
        Heating Elements - 5.5kW, Motor - 15.3A
Furnace (8/6kW - 240/208V)
AC (1645/2188VA - 240/208V)
Blower - ½HP blower (240/208V)

Exercise Room and Spa - 650SF
15 - Duplex receptacles
10 - 2 x 4 fluorescent light fixtures (1.65A per ballast - 120V)
Furnace (13.3/10kW – 240/208V)
AC (2743/3648VA – 240/208V)
Blower - ¾HP blower (240/208V)
 2 - Jacuzzi heated air tubs
        3.3/2.5kW single-phase heaters (240/208V)
        ¾ HP circulatory pump motors (120V)

Swimming Pool
5HP pump motor (208V - 3φ)
2HP circulatory motor (208V - 3φ)
8 - underwater light fixtures (1.23A - 120V)

Outside Lighting
13 - Metal halide poles mounted fixtures (2.48/3.3A – 240/208V)
 3 - Signs (2kW - 120V-flurorescent)

Calculate the (3φ) house load and the 54 unit apartment complex service and neutral loads. Determine the size service required to supply both loads.

Calculate the individual (3φ) service and neutral loads per apartment units and each (1φ) individual unit. Determine the size service, service and neutral conductors (75°C - copper) needed per apartment units and each individual unit. Calculate the individual (1φ) service and neutral loads per unit.

Answers to question No. 87 consist of 52 pages (pages 74 – 126)

HOUSE LOAD - THREE-PHASE LOAD CALCULATIONS AT 208/120V

For "House Load" reference format see **(Worksheet G** - Volume 4**)** STANDARD LOAD CALCULATION FOR NONDWELING BUILDING (COMMERICIAL and INDUSTRIAL).

Note - Although the option of using a combination of main overcurrent devices is listed in this situation and hereafter, the use of a single main overcurrent device is only considered for sizing feeder or service conductors.

HOUSE LOAD - 208/120V 3-phase (House Load calculation covers pages 74 – 79)

1. GENERAL LIGHTING or ACTUAL LIGHTING LOADS

General Lighting Load <NEC References - 220.12 and Table 220.12>

Office - 740SF x 3.5VA = 2590VA

    A. General Lighting Load = ~~2590VA~~ (Omit General Lighting Load - Actual Lighting Load Larger)

Actual Lighting Load

Type Fixture	VA rating	No. of Fixtures		TOTAL VA
Office				
Fluorescent	120V x 1.65A x	12	=	2376
Recessed Cans	100W(VA) x 5		=	500
B. Actual Lighting Load			=	2876
	2876VA x 1.25		=	3595VA

1.	LINE LOAD	NEUTRAL LOAD	
		Permitted Reduction	Prohibited Reduction
	3595VA	500VA	2376VA

2. OTHER LIGHTING LOADS

A. Sign/Outline (S/O) Lighting <NEC References - 220.12(F) and 600.5(A)>

Signs - 2000VA x 3 = 6000VA

Total (Sign/Outline Lighting) = 6000VA

B. Outside Lighting <NEC Reference - 220.18(B)>

Metal Halide - 208V x 3.3A x 13 = 8923.2VA

Total (Outside Lighting) = 8923.2VA

C. NA

D. Track Lighting <NEC Reference - 220.43(B)> (Voltage rating -120V)

(8' ÷ 2') x 150VA = 600VA

E. Miscellaneous (Write-ins. List individual voltage rating of each lighting load.)

1. Underwater light fixtures - 120V x 1.23A x 8 = 1180.8VA
2. Laundromat, Exercise Room and Spa (Fluorescent) - 120V x 1.65A x 16 = 3168VA

Total (Miscellaneous) = 4348.8VA

F. Other Lighting (LINE and NEUTRAL) Loads Total [Add lines (A.) - (E.)]

Other Lighting Loads Total = 19,872VA

TOTAL = 19,872VA x 1.25 = 24,840VA

2.	LINE LOAD	NEUTRAL LOAD	
		(120V or 277V)	
		Permitted Reduction	Prohibited Reduction
	24,840VA	1780.8VA (D. and E1.)	9168VA (A. and E2.)

3. RECEPTACLE LOADS

A. Non-continuous duty (Other Than Bank or Office Building) <NEC 220.14(I)>

Laundromat (5), Exercise Room and Spa (15) - 20 x 1800VA = 3600VA

B. Non-continuous duty (Office Building) <NEC 220.14(K)>

Office

(1) 10 x 180VA = 1800VA (larger)

(2) 740SF x 1VA = 740VA

C. NA

D. Total Receptacle (LINE and NEUTRAL) Load

Total Receptacle loads - 3600VA + 1800VA = 5400VA
LINE = 5400VA
NEUTRAL = 5400VA

3.   LINE LOAD

	NEUTRAL LOAD	
	Permitted Reduction	Prohibited Reduction
5400VA	5400VA	0

4. KITCHEN EQUIPMENT - NA

5. SPECIFIC LOADS (Appliances, Computer Equipment, Office equipment, etc.)

Type Load	Calculation	
208V		
Copier	208V x 13.7A =	2849.6VA
Dryer (Heating Elements) (6)	5500W x 6 =	33,000.0VA
Jacuzzi Heater (2)	2500W x 2 =	5000.0VA
Water Heater	3380W =	3380.0VA
		44,229.6VA
120V		
Desktop computers (4)	120V x 1.3A x 4 =	624.0VA
Laser printers (2)	120V x 3.8A x 2 =	912.0VA
		1536.0VA

TOTAL = 45,765.6VA

4. LINE LOAD

	NEUTRAL LOAD	
	Permitted Reduction	Prohibited Reduction
45,765.6VA	0	1536VA

6. MOTOR LOADS <NEC Reference - 220.50>

A. Continuous Duty

Motor Load	Calculation		

120V

¾HP Circulatory Pump (2)	120V x 13.8A* x 2	=	3312.00VA

208V-3φ

5HP Swimming Pool Pump	208V x 16.7A** x 1.732	=	6016.28VA
2HP Swimming Pool Pump	208V x 7.5A** x 1.732	=	2701.92VA
			12,030.20VA

*Table 430.248   **Table 430.250

B. Non-Continuous Duty

Motor Load	Calculation		

120V

Soft drink machine	120V x 10.4A	=	1248.00VA

208V- 3φ

Dryers (6)	208V x 15.3A x 1.732 x 6	= 33,071.50VA
Washers (12)	208V x 13A x 1.732 x 12	= 56,199.94VA
		89,271.44VA

C. and D. - NA

E. Total Motor Loads – [LINE LOAD - Add lines (A.) and (B.)] – [NEUTRAL LOAD - Total motor loads with neutral connections (120V)]

TOTAL = 102,549.64VA

6.	LINE LOAD	NEUTRAL LOAD	
		Permitted Reduction	Prohibited Reduction
	102,549.64VA	4560VA	0

7. MEDICAL EQUIPMENT - NA

8. INDUSTRIAL EQUIPMENT - NA

9. HEATING and AIR-CONDITIONING (AC) EQUIPMENT <NEC References - 220.50, 220.51, 220.60, 430.6(A)(1) and 440.6(A)>

ELECTRICAL HEATING UNITS

Heating Unit	Calculation
Furnaces - 10kW (2)	10,000W x 2 = 20,000W (VA)
Furnace - 6kW	= 6000W (VA)

TOTAL HEAT = 26,000VA

AIR-CONDITIONING (AC) UNITS

AC Unit	Calculation
Units - 3648VA (2)	3648VA x 2 = 7296VA
Unit - 2188VA	= 2188VA

TOTAL AC = 9484VA

Line Load = 26,000VA + (4284.8VA) = 30,284.8VA
            (Largest Load)   (Air Handlers [Blowers])

¾ HP - 208V x 7.6A* x 2 = 3161.6VA   *Table 430.248
½ HP - 208V x 5.4A* = 1123.2VA
                        4284.8VA

9.	LINE LOAD		NEUTRAL LOAD	
			Permitted Reduction	Prohibited Reduction
	30,284.8VA		0	0

10. LARGEST MOTOR

5HP Swimming Pool Motor (motor with highest current per NEC 430.17 and 440.7)

6106.28VA x 25 percent (.25) = 1504.1VA

10.	LINE LOAD		NEUTRAL LOAD	
			Permitted Reduction	Prohibited Reduction
	1504.1VA		0	0

TOTAL DEMAND LOAD (LINE and NEUTRAL) (List each computed line and neutral loads below and total lines 1. – 10.)

	LINE LOAD	NEUTRAL LOAD	
		Permitted Reduction	Prohibited Reduction
1. General Lighting	3595.00VA	500.0VA	2376.0VA
2. Other Lighting Loads	24,840.00VA	1780.8VA	9168.0VA
3. Receptacle Loads	5400.00VA	5400.0VA	0
4. Kitchen Equipment	--	--	--
5. Specific Loads	45,765.60VA	0	1536.0VA
6. Motor Loads	102,549.64VA	4560.0VA	0
7. Medical Equipment	--	--	--
8. Industrial Equipment	--	--	--
9. Heating and AC Equip.	30,284.80VA	0	0
10. Largest Motor	1504.10VA	0	0
TOTAL =	213,939.14VA	12,240.8VA	13,080.0VA

BUILDING'S OPERATING LINE VOLTAGE - 208V(3ϕ)
(Given operating voltage or as determined per test examination)

CALCULATE MINIMUM LINE and NEUTRAL LOADS
(Divide Total Demand Load **[VA]** by operating line voltage **[V]**)

LINE LOAD = 213,939.14VA / 208V x 1.732 = 593.85A

NEUTRAL LOAD
    Permitted   = 12,240.8VA / 208V x 1.732 = 33.98A
    Prohibited   = 13,080.0VA / 208V x 1.732 = 36.31A

               Total Neutral Load = 70.29A

SIZE SERVICE (Size of service based on the calculated LINE LOAD)

SIZE SERVICE REQUIRED (minimum) 600A
(Single rating or combination - Main overcurrent device(s) to total 600A)

The house service load is 593.85A (3ϕ) and the neutral load is 70.29A.

> ## THREE-PHASE LOAD CALCULATIONS PER APARTMENT COMPLEX AT 208/120V

For "Multifamily Dwelling" reference format see **Worksheet C** (Volume 4) – STANDARD LOAD CALCULATIONS FOR MULTIFAMILY DWELLING.

<u>54 units - 208/120V 3-phase</u> (54 units load calculation covers pages 80 – 86)

1. General Lighting and Receptacle Loads <NEC 220.12, Table 220.12, 220.14(J) and 220.42> (Open porches, garages, unused or unfinished spaces not adaptable for future use not included.)

Floor Plan 1 -   650SF x <u>15</u> x 3VA = 29,250VA
Floor Plan 2 -   900SF x <u>17</u> x 3VA = 45,900VA
Floor Plan 3 - 1200SF x <u>12</u> x 3VA = 43,200VA
Floor Plan 4 - 1400SF x <u>10</u> x 3VA = <u>42,000VA</u>

Total (Floor Plans) = 160,350VA

2. Small-Appliance Circuit Load (Portable) <NEC 220.52(A)> (Minimum of two circuits required.)

1500VA x 2 x <u>54</u> = 162,000VA

3. Laundry Circuit Load <NEC 220.52(B)>

1500VA x <u>39</u>[1] = 58,500VA
[1]650SF units **(15)** not equipped with washer/dryer connections.

TOTAL (Lines 1. - 3.) =   380,850VA
(If Total VA is less than or equal to 120,000VA, step c. is not required)

APPLY DEMAND FACTORS <NEC and Table 220.42>

a. First 3000VA of above TOTAL (At 100%)   =   3000.0VA

b. 117,000VA   x .35   =   40,950.0VA
   (Total VA – 3001VA up to 117,000VA)

c. 260,850VA   x .25   =   65,212.5VA
   (Remainder of TOTAL exceeding 120,000VA)*

   *(380,850VA – 120,000 VA = 260,850VA)

TOTAL (Lines a. - c.)   = 109,162.5VA

GENERAL LIGHTING and RECEPTACLE, SMALL-APPLIANCE and LAUNDRY LOADS

1. - 3. <u>LINE LOAD</u>   <u>NEUTRAL LOAD</u>

    109,162.5VA   109,162.5VA

4. Appliance Loads (Fastened-In-Place) <NEC 220.53> (Use nameplate rating of each appliance. Electric ranges, dryers, space-heating equipment or air-conditioning equipment not included.)

120V Appliances	VA	No. of Appliances	Total VA
1. Dishwashers	900	32	28,800
2. Dishwashers	1250	22	27,500
3. Disposals	600	32	19,200
4. Disposals	900	22	19,800
5. Trash compactors	780	22	17,160
6. Microwaves	1000	15	15,000
7. Microwaves	1200	17	20,400
8. Microwaves	1600	12	19,200
9. Microwaves	1800	10	18,000
10. Heat-Vent-Lights	1300	15	19,500
11. Heat-Vent-Lights	1600	29	46,400
12. Heat-Vent-Lights	1750	10	<u>17,500</u>

              (Total 120V Appliances) 268,460

208V Appliances	VA	No. Appliances	Total VA
1. Water Heaters	3880	54	<u>182,520</u>

              (Total 208V Appliances) 182,520

            APPLIANCES TOTAL   = 450,980VA

APPLY DEMAND FACTOR (Applicable, when number of above appliances exceeds four (4) or more.)

        (Appliances Total) <u>450,980VA</u> x .75 = 338,235VA

APPLIANCE LOADS

4.   <u>LINE LOAD</u>   <u>NEUTRAL LOAD</u> (Refer to condition)

    338,235VA   201,345VA

Condition
Because the water heaters are 208 volts line-to-line loads they will not contribute to the neutral load as the other appliances and therefore must not be included. As a result, the NEUTRAL LOAD is equal to

$$450,980VA - 182,520 \times .75 = 201,345VA$$

5. Clothes Dryers[2] (208/120V-1ϕ) <NEC and Table 220.54> (Use 5000W [VA] or nameplate rating, whichever is larger.)

Dryer Load determined per single-phase dryers supplied by 3-phase 4W Feeder or Service

[2]650SF units (15) not equipped with washer/dryer connections.

Number of dryers = 39 (17+12+10)

$\frac{39}{3}$ = 13 (number of dryers per phase)

13 x 2 = 26 (number of dryers)

Based on 26 dryers,

the demand factor (%) = 35% − .5% x (26 − 23) = 33.5% (.335)

5000VA x 26 x .335 = 43,550VA (Demand Load for 2 phases)

$\frac{43,550VA}{2 \text{ (phases)}}$ = 21,775VA [Demand Load per phase (1ϕ)]

21,775VA [Demand Load per phase (1ϕ)] x 3 = 65,325VA [3-phase (3ϕ) Demand Load]

CLOTHES DRYERS

5.　　LINE LOAD　　　　　　　　NEUTRAL LOAD [NEC 220.61(C)(1)]

65,325VA　　　　　　　　　　65,325VA

6. Electric Ranges and Other Cooking Appliances <NEC and Table 220.55>

LINE LOAD – 208/120V (Ranges and WMO) and 208V (Cooktop)

Number of ranges and cooking appliances = 76 (15+17+24+20). Refer to question No. 70. for additional details.

The ranges and cooking appliances total kW rating as determined:

$$
\begin{array}{lll}
\text{Ranges -} & 15 \times 6100W = & 91,500W \\
\text{Ranges -} & 17 \times 7800W = & 132,600W \\
\text{Cooktops -} & 12 \times 6300W = & 75,600W \\
\text{WMOs -} & 12 \times 5400W = & 64,800W \\
\text{Cooktops -} & 10 \times 6700W = & 67,000W \\
\text{WMOs -} & \underline{10} \times 5800W = & \underline{58,000W} \\
& 76 & 489,500W \ (489.5kW)
\end{array}
$$

Per Column B - Apply demand factor and determine demand load.

76 Cooking Appliances = .16
489.5kW x .16 = 78.32kW (78,320W)

Per Column C - Determine maximum demand. Use lowest value of Columns B and C.

Max Demand for 76 Ranges = 25kW + (.75kW x 76) = 82kW (82,000W)

RANGE/COOKING APPLIANCES

6.	LINE LOAD	NEUTRAL LOAD [NEC 220.61(B)(1)] (Refer to condition)
	78,320VA	62,676VA

Condition
Because the cooktops are 208 volts line-to-line loads they will not contribute to the neutral load as the cooking appliances and therefore must not be included. As a result, the range/cooking appliances LINE LOAD must be recalculated omitting the cooktops to determine the NEUTRAL LOAD.

NEUTRAL LOAD – 208/120V (Ranges and WMO)

Number of ranges and WMOs (with neutral connections) = 54 (15+17+12+10)

The ranges and WMOs total kW rating = 346,900W (346.9kW)

Per Column B (lowest value)   Per Column C
54 Cooking Appliances = .18 (demand factor) Max Demand for 76 Ranges
346.9kW x .18 = 62.442kW (62,442W)   25kW + (.75kW x 54) = 65.5kW (65,500W)

$$\frac{51kW}{2 \text{ (phases)}} = 25.5kW \text{ [Demand Load per phase (1}\phi\text{)]}$$

7. Heating and Air-Conditioning (AC) Equipment <NEC References - 220.50, 220.51, 220.60, Table 430. 248 and 440.6(A)> (Include VA rating of air handler [blower].)

Heating Load (larger)

Electric Heat
10kW (10,000VA) x <u>15</u> = 150,000VA
12kW (120,00VA) x <u>17</u> = 204,000VA
15kW (150,00VA) x <u>12</u> = 180,000VA
18kW (180,00VA) x <u>10</u> = <u>180,000VA</u>
714,000VA

AC Load

3648VA x <u>15</u> = 54,720VA
4560VA x <u>17</u> = 77,520VA
5472VA x <u>12</u> = 65,664VA
6384VA x <u>10</u> = <u>63,840VA</u>
261,744VA

Electric Heat   Blowers (1)   Blowers (2)   Blowers (3)
Line Load = <u>714,000VA</u> + <u>1044VA</u> x <u>15</u> + <u>1176VA</u> x <u>17</u> + <u>1320VA</u> x <u>12</u>
Blowers (4)
+ <u>1416VA</u> x <u>10</u> = <u>779,652VA</u>

HEATING and AC

7.   <u>LINE LOAD</u>   <u>NEUTRAL LOAD</u>

779,652VA   0

8. Largest Motor <NEC 430.17, 430.24, 440.7 and 440.33> (Use motor with highest full-load current [FLC] regardless of voltage rating)

Largest Motor (LM) = 900VA (7.5A) disposal rated for 120 volts

<u>900VA</u> x .25 = <u>225VA</u>
(LM)

LARGEST MOTOR

8.   <u>LINE LOAD</u>   <u>NEUTRAL LOAD (120V motors only)</u>
225VA   225VA

TOTAL DEMAND LOAD (LINE and NEUTRAL) (Add lines 1. – 8.)

	LINE LOAD	NEUTRAL LOAD
1. - 3.	109,162.5VA	109,162.5VA
4.	338,235.0VA	201,345.0VA
5.	65,325.0VA	65,325.0VA
6.	78,320.0VA	62,442.0VA
7.	779,652.0VA	0
8.	225.0VA	225.0VA
	1,370,919.5VA	438,499.5VA

DWELLING'S OPERATING LINE VOLTAGE - 208V(3φ)
(Given operating voltage or as determined per test examination)

CALCULATE MINIMUM LINE and NEUTRAL LOADS
(Divide Total Demand Load [VA] by operating line voltage [V])

LINE LOAD = 1,370,919.5VA / 208V x 1.732 = 3805.4A

NEUTRAL LOAD
Because the DRYER (65,325VA) and RANGE/COOKING APPLIANCE (62,442VA) loads are fed from circuits which are prohibited per NEC 220.61(C)(1), the reduction of both neutral loads per NEC 220.61(B)(1) cannot be applied. Combined DRYER and RANGE/COOKING APPLIANCE neutral loads = 127,767VA.

Permitted Reduction

438,499.5VA – 127,767VA = 310,732.5VA (Permitted)

310,732.5VA / 208V x 1.732 = ~~862.53A~~* 663.77A

*Where the neutral load exceeds 200A, NEC 220.61(B)(2) permits the load to be reduced by 70 percent. Complete the following to determine permitted Neutral Demand Load.

(1) 862.53   A – 200A = 662.53   A x .70 = 463.77A
        (Neutral Load)*   (Remainder)   (Permitted Reduction)

(2) 463.77   A + 200A = 663.77   A
        (Permitted Reduction)   (Permitted Neutral Load)

Prohibited Reduction

127,767VA / 208V x 1.732 = 354.66A

NEUTRAL LOAD = 663.77A + 354.66A = 1018.43A

SIZE SERVICE (Size of service based on the calculated **LINE LOAD**)

SIZE SERVICE REQUIRED (minimum) <u>4000A</u>
(Single rating or combination - Main overcurrent device(s) to total 4000A)

The overall apartment complex service load is 3805.4A (3φ) and the neutral load is 1018.43A.

THREE-PHASE LOAD CALCULATIONS PER APARTMENT UNITS AT 208/120V

For "Multifamily Dwelling" reference format see **Worksheet C** (Volume 4) – STANDARD LOAD CALCULATIONS FOR MULTIFAMILY DWELLING.

<u>**650SF units (15) - 208/120V 3-phase**</u> (650SF units load calculation covers pages 86 – 91)

1. General Lighting and Receptacle Loads <NEC 220.12, Table 220.12, 220.14(J) and 220.42> (Open porches, garages, unused or unfinished spaces not adaptable for future use not included.)

650SF x 15 x 3VA = 29,250VA

2. Small-Appliance Circuit Load (Portable) <NEC 220.52(A)> (Minimum of two circuits required.)

1500VA x 2 x 15 = 45,000VA

3. Laundry Circuit Load <NEC 220.52(B)[1]> = 0

[1]650SF units **(15)** not equipped with washer/dryer connections.

TOTAL (Lines 1. - 3.) = 74,250VA
(If Total VA is less than or equal to 120,000VA, step c. is not required)

APPLY DEMAND FACTORS <NEC and Table 220.42>

a. First 3000VA of above TOTAL (At 100%) = 3000.0VA

b. 71,250VA   x .35   = 24,937.5VA
    (Total VA – 3001VA up to 117,000VA)

c.   0   x .25   =   0
    (Remainder of TOTAL exceeding 120,000VA)*

TOTAL (Lines a. - c.)   = 27,937.5VA

GENERAL LIGHTING and RECEPTACLE, SMALL-APPLIANCE and LAUNDRY LOADS

1. - 3.	LINE LOAD	NEUTRAL LOAD
	27,937.5VA	27,937.5VA

4. Appliance Loads (Fastened-In-Place) <NEC 220.53> (Use nameplate rating of each appliance. Electric ranges, dryers, space-heating equipment or air-conditioning equipment not included.)

120V Appliances	VA	No. of Appliances	Total VA
1. Dishwashers	900	15	13,500
2. Disposals	600	15	9000
3. Microwaves	1000	15	15,000
4. Heat-Vent-Lights	1300	15	19,500
		(Total 120V Appliances)	57,000

208V Appliances	VA	No. Appliances	Total VA
1. Water Heaters	3380	15	50,700
		(Total 208V Appliances)	50,700

APPLIANCES TOTAL   = 107,700VA

APPLY DEMAND FACTOR (Applicable, when number of above appliances exceeds four (4) or more.)

(Appliances Total) 107,700VA x .75 = 80,775VA

APPLIANCE LOADS

4.	LINE LOAD	NEUTRAL LOAD (Refer to condition)
	80,775VA	42,750VA

Condition

Because the water heaters are 208 volts line-to-line loads they will not contribute to the neutral load as the other appliances and therefore must not be included. As a result, the NEUTRAL LOAD is equal to

107,700VA − 50,700VA x .75 = 42,750VA

5. Clothes Dryers[2] (208/120V-1ϕ) <NEC and Table 220.54> (Use 5000W [VA] or nameplate rating, whichever is larger.)

[2]650SF units **(15)** not equipped with washer/dryer connections.

CLOTHES DRYERS

5. <u>LINE LOAD</u>   <u>NEUTRAL LOAD</u>

    0   0

6. Electric Ranges and Other Cooking Appliances (208/120V-1ϕ) <NEC and Table 220.55>

Range Load determined per single-phase ranges supplied by 3-phase 4W Feeder or Service

Number of ranges = 15 [@6.1kW (6100VA)]

$\dfrac{15}{3}$ = 5(number of ranges per phase)

5 x 2 = 10 (number of ranges)

Per Column B to Table 220.55

Based on 10 ranges, demand factor (%) = 34(.34)
the demand load = 6100VA x 10 x .34 = 20,740VA (Demand Load for 2 phases)

$\dfrac{20{,}740VA}{2\ (phases)}$ =   10,370VA [Demand Load per phase (1ϕ)]

10,370VA [Demand Load per phase (1ϕ)] x 3 = 31,110VA [3-phase (3ϕ) Demand Load]

Per Column C to Table 220.55

Based on 10 ranges Maximum Demand
Maximum Demand (Demand Load for 2 phases) = 25kW (25,000VA)

$\dfrac{25{,}000VA}{2\ (phases)}$ = 12,500VA [Demand Load per phase (1ϕ)]

12,500VA [Demand Load per phase (1ϕ)] x 3 = 37,500VA [3-phase (3ϕ) Demand Load]

Compared to the Maximum Demand of Column C (37,500VA) - Use 31,110VA as the demand load for range appliances.

RANGE/COOKING APPLIANCES

6. <u>LINE LOAD</u>   <u>NEUTRAL LOAD [NEC 220.61(C)(1)]</u>

    31,110VA   10,370VA

7. Heating and Air-Conditioning (AC) Equipment <NEC References - 220.50, 220.51, 220.60, Table 430. 248 and 440.6(A)> (Include VA rating of air handler [blower].)

Heating Load (larger)

Electric Heat

$$10kW\ (10,000VA) \times \underline{15} = 150,000VA$$

AC Load

$$3648VA \times \underline{15} = 54,720VA$$

$$\text{Line Load} = \underset{\text{Electric Heat}}{\underline{150,000VA}} + \underset{\text{Blowers}}{\underline{1044VA}} \times \underline{15} = \underline{165,660VA}$$

HEATING and AC

7.	LINE LOAD	NEUTRAL LOAD
	165,660VA	0

8. Largest Motor <NEC 430.17, 430.24, 440.7 and 440.33> (Use motor with highest full-load current [FLC] regardless of voltage rating.)

Largest Motor (LM) = 1044VA (5.02A) disposal rated for 208 volts

$$\underset{\text{(LM)}}{\underline{1044VA}} \times .25 = \underline{261VA}$$

LARGEST MOTOR

8.	LINE LOAD	NEUTRAL LOAD (120V motors only)
	261VA	0

TOTAL DEMAND LOAD (LINE and NEUTRAL) (Add lines 1. – 8.)

	LINE LOAD	NEUTRAL LOAD
1. - 3.	27,937.5VA	27,937.5VA
4.	80,775.0VA	42,750.0VA
5.	0	0
6.	31,110.0VA	31,110.0VA
7.	165,660.0VA	0
8.	261.0VA	0
	305,743.5VA	101,797.5VA

DWELLING'S OPERATING LINE VOLTAGE - <u>208V</u>(3ϕ)
(Given operating voltage or as determined per test examination)

CALCULATE MINIMUM LINE and **NEUTRAL LOADS**
(Divide Total Demand Load **[VA]** by operating line voltage **[V]**)

LINE LOAD = <u>305,743.5</u>VA / <u>208V</u> x 1.732 = <u>848.68</u>A

NEUTRAL LOAD
Because the RANGE (31,110VA) loads are fed from circuits which are prohibited per NEC 220.61(C)(1), the reduction of the neutral load per NEC 220.61(B)(1) & (2) cannot be applied.

Permitted Reduction

      101,797.5VA – 31,110VA = 70,687.5VA (Permitted)

      <u>70,687.5</u>VA / <u>208V</u> x 1.732 = <u>196.21</u>A (less than 200 amps – 70 percent reduction not permitted)

Prohibited Reduction

      <u>31,110</u>VA / <u>208V</u> x 1.732 = <u>86.36</u>A

NEUTRAL LOAD = <u>196.21</u>A + <u>86.36</u>A = <u>282.57</u>A

SIZE SERVICE (Size of service based on the calculated **LINE LOAD**)

        SIZE SERVICE REQUIRED (minimum) <u>1000A</u>
      (Single rating or combination - Main overcurrent device(s) to total 1000A)

The conductors or combinations selected for this installation are only limited for this question. Conductor ampacity derating is not applied in sizing the service and neutral conductors.

SIZING FEEDER/SERVICE CONDUCTORS

Using NEC 310.10(H)(1) and Table 310.15(B)(16), as a minimum based on NEC 240.4(C), use three 400 kcmil copper conductors (75°C) per phase which have an individual ampacity of 335A, as the service conductors.

      FEEDER/SERVICE CONDUCTORS <u>3-400 kcmil copper conductors per line</u>

SIZING NEUTRAL CONDUCTOR(S)

Based on the calculated neutral load and the use of parallel service conductors, the neutral conductors must also be installed in parallel per NEC 310.10(H)(1). Because the service conductors will consist of three parallel conductors per phase, the neutral conductors must also consist of three parallel conductors based on the calculated neutral load being divided by three which results to 94.2A (282.57A / 3). Per Table 310.15(B)(16), at 75°C, a 3 AWG copper conductor which has a rated ampacity of 100A will satisfy the calculated neutral load being distributed amongst three such conductors. However, because NEC 310.10(H)(1) only allows a 1/0 AWG conductor or larger to be installed in parallel, three 1/0 AWG copper conductors must be used instead as the parallel neutral conductors.

In total circular mils, the three 1/0 AWG parallel neutral conductors will exceed the required 3/0 AWG grounding electrode conductor, when considered. The size of the grounding electrode conductor is based on the equivalent area of the parallel service conductors [1200 kcmil (400kcmil x 3)] per Table 250.66 which is over 1100 kcmil copper.

NEUTRAL CONDUCTOR(S) <u>3-1/0 AWG copper conductors</u>

See question Nos. 1. - 8. of Article 250 for further reference when the grounded (neutral) conductor is not used as a circuit conductor.

<u>900SF units (17) - 208/120V 3-phase</u> (900SF units load calculation covers pages 91 – 96)

1. General Lighting and Receptacle Loads <NEC 220.12, Table 220.12, 220.14(J) and 220.42> (Open porches, garages, unused or unfinished spaces not adaptable for future use not included.)

$$900SF \times 17 \times 3VA = 45,900VA$$

2. Small-Appliance Circuit Load (Portable) <NEC 220.52(A)> (Minimum of two circuits required.)

$$1500VA \times 2 \times 17 = 51,000VA$$

3. Laundry Circuit Load <NEC 220.52(B)>

$$1500VA \times 17 = 25,500VA$$

TOTAL (Lines 1. - 3.) = 122,400VA
(If Total VA is less than or equal to 120,000VA, step c. is not required)

APPLY DEMAND FACTORS <NEC and Table 220.42>

a. First 3000VA of above TOTAL (At 100%) =   3000VA

b. 117,000VA   x .35   =   40,950VA
   (Total VA – 3001VA up to 117,000VA)

c. 2400VA   x .25   =   600VA
  (Remainder of TOTAL exceeding 120,000VA)*

TOTAL (Lines a. - c.)   =   44,550VA

GENERAL LIGHTING and RECEPTACLE, SMALL-APPLIANCE and LAUNDRY LOADS

1. - 3. <u>LINE LOAD</u>   <u>NEUTRAL LOAD</u>

    44,550VA   44,550VA

4. Appliance Loads (Fastened-In-Place) <NEC 220.53> (Use nameplate rating of each appliance. Electric ranges, dryers, space-heating equipment or air-conditioning equipment not included.)

120V Appliances	**VA**	**No. of Appliances**	**Total VA**
1. Dishwasher	900	17	15,300
2. Disposal	600	17	10,200
3. Microwave	1200	17	20,400
4. Heat-Vent-Light	1600	17	27,200
		(Total 120V Appliances)	73,100

208V Appliances	**VA**	**No. Appliances**	**Total VA**
1. Water Heater	3380	17	57,460
		(Total 208V Appliances)	57,460

          APPLIANCES TOTAL   = 130,560VA

APPLY DEMAND FACTOR (Applicable, when number of above appliances exceeds four (4) or more.)

        (Appliances Total) <u>130,560VA</u> x .75 = 97,920VA

APPLIANCE LOADS

4.   <u>LINE LOAD</u>   <u>NEUTRAL LOAD</u> (Refer to condition)

    97,920VA   54,825VA

<u>Condition</u>
Because the water heaters are 208 volts line-to-line loads they contribute to the neutral load as the other appliances and therefore must not be included. As a result, the NEUTRAL LOAD is equal to

        130,560VA – 57,460VA x .75 = 54,825VA

5. Clothes Dryers (208/120V-1φ) <NEC and Table 220.54> (Use 5000W [VA] on nameplate rating, whichever is larger.)

    Dryer Load determined per single-phase dryers supplied by 3-phase 4W Feeder or Service

    Number of dryers = 17 $\frac{17}{3}$ = 5.6 6 (number of dryers per phase)

$6 \times 2 = 12$ (number of dryers)

Based on 12 dryers,

the demand factor (%) = 47% - 1% x (12 − 11) = 46%(.46)

5000VA x 12 x .46 = 27,600VA (Demand Load for 2 phases)

$$\frac{27,600VA}{2 \text{ (phases)}} = 13,800VA \text{ [Demand Load per phase (1}\phi\text{)]}$$

13,800VA [Demand Load per phase (1ϕ)] x 3 = 41,400VA [3-phase (3ϕ) Demand Load]

CLOTHES DRYERS

5.   <u>LINE LOAD</u>   <u>NEUTRAL LOAD [NEC 220.61(C)(1)]</u>

41,400VA   13,800VA

6. Electric Ranges and Other Cooking Appliances (208/120V-1ϕ) <NEC and Table 220.55>

Range Load determined per single-phase ranges supplied by 3-phase 4W Feeder or Service

Number of ranges = 17 [@7.8kW (7800VA)]

$$\frac{17}{3} = \text{5.6 } 6 \text{ (number of ranges per phase)}$$

$6 \times 2 = 12$ (number of ranges)

Per Column B to Table 220.55

Based on 12 ranges, demand factor (%) = 32(.32)
the demand load = 7800VA x 12 x .32 = 29,952VA (Demand Load for 2 phases)

$$\frac{29,952VA}{2 \text{ (phases)}} = 14,976VA \text{ [Demand Load per phase (1}\phi\text{)]}$$

14,976VA [Demand Load per phase (1ϕ)] x 3 = 44,928VA [3-phase (3ϕ) Demand Load]

Per Column C to Table 220.55

Based on 12 ranges, Maximum Demand
(Demand Load for 2 phases) = 27kW (27,000VA)

$$\frac{27{,}000VA}{2 \text{ (phases)}} = 13{,}500VA \text{ [Demand Load per phase (1}\phi\text{)]}$$

13,500VA [Demand Load per phase (1φ)] x 3 = 40,500VA [3-phase (3φ) Demand Load]

Compared to the Maximum Demand of Column C (40,500VA) - Use 40,500VA as the demand load for range appliances.

RANGE/COOKING APPLIANCES

6.	LINE LOAD	NEUTRAL LOAD [NEC 220.61(C)(1)]
	40,500VA	40,500VA

7. Heating and Air-Conditioning (AC) Equipment <NEC References - 220.50, 220.51, 220.60, Table 430. 248 and 440.6(A)> (Include VA rating of air handler [blower].)

Heating Load

Electric Heat

12kW (12,000VA) x 17 = 204,000VA

AC Load

4560VA x 17 = 77,520VA

Heating load is larger.

Electric Heat   Blowers
Line Load = 204,000VA + 1176VA x 17 = 223,992VA

HEATING and AC

7.	LINE LOAD	NEUTRAL LOAD
	223,992VA	0

8. Largest Motor <NEC 430.17, 430.24, 440.7 and 440.33> (Use motor with highest full-load current [FLC] regardless of voltage rating.)

Largest Motor (LM) = 1176VA (5.65A) blower rated for 208 volts

$\underline{1176VA}$ x .25 = $\underline{294VA}$
(LM)

LARGEST MOTOR

8. <u>LINE LOAD</u>   <u>NEUTRAL LOAD (120V motors only)</u>

    294VA   0

TOTAL DEMAND LOAD (LINE and NEUTRAL) (Add lines 1. – 8.)

	LINE LOAD	NEUTRAL LOAD
1. - 3.	44,550VA	44,550VA
4.	97,920VA	54,825VA
5.	41,400VA	13,800VA
6.	40,500VA	13,500VA
7.	223,992VA	0
8.	294VA	0
	448,656VA	126,675VA

DWELLING'S OPERATING LINE VOLTAGE - $\underline{208V}$(3φ)
(Given operating voltage or as determined per test examination)

CALCULATE MINIMUM LINE and NEUTRAL LOADS
(Divide Total Demand Load **[VA]** by operating line voltage **[V]**)

LINE LOAD = $\underline{448,656}$VA / $\underline{208}$V x 1.732 = $\underline{1245.38A}$

NEUTRAL LOAD
Because the DRYER (13,800VA) and RANGE (13,500VA) loads are fed from circuits which are prohibited per NEC 260.61(C)(1), the reduction of both neutral loads per NEC 260.61(B) cannot be applied. Combined DRYER and RANGE neutral loads = 27,300VA.

Permitted Reduction

126,675VA – 27,300VA = 99,375VA (Permitted)

$\underline{99,375}$VA / $\underline{208}$V x 1.732 = ~~275.85A~~* 253.1A

*Where the neutral load exceeds 200A, NEC 220.61(B)(2) permits the load to be reduced by 70 percent. Complete the following to determine permitted Neutral Demand Load.

(1) $\underline{275.85}$ A – 200A = $\underline{75.85}$ A x .70 = $\underline{53.1}$ A
    (Neutral Load)*   (Remainder)   (Permitted Reduction)

(2) $\underline{53.1}$ _____ A + 200A = $\underline{253.1}$ _____ A
     (Permitted Reduction)   (Permitted Neutral Load)

Prohibited Reduction

    $\underline{27,300}$VA / $\underline{208V}$ x 1.732 = $\underline{75.78A}$

NEUTRAL LOAD = $\underline{253.1}$A + $\underline{75.78}$A = $\underline{328.88}$A

SIZE SERVICE (Size of service based on the calculated **LINE LOAD**)

SIZE SERVICE REQUIRED (minimum) $\underline{1600A}$
(Single rating or combination - Main overcurrent device(s) to total **1600A**)

The conductors or combinations selected for this installation are only limited for this question. Conductor ampacity derating is not applied in sizing the service and neutral conductors.

SIZING FEEDER/SERVICE CONDUCTORS

Using NEC 310.10(H)(1) and Table 310.15(B)(16), as a minimum based on NEC 240.4(C), use four 600 kcmil copper conductors (75°C) per phase which have an individual ampacity of 420A, as the service conductors.

FEEDER/SERVICE CONDUCTORS $\underline{\text{4-600 kcmil copper conductors per line}}$

SIZING NEUTRAL CONDUCTOR(S)

Based on the calculated neutral load and the use of parallel service conductors, the neutral conductors must also be installed in parallel per NEC 310.10(H)(1). Because the service conductors will consist of four parallel conductors per phase, the neutral conductors must also consist of four parallel conductors based on the calculated neutral load being divided by four which results to 82.22A (328.88A / 4). Per Table 310.15(B)(16), at 75°C, a 4 AWG copper conductor which has a rated ampacity of 85A will satisfy the calculated neutral load being distributed amongst four such conductors. However, because NEC 310.10(H)(1) only allows a 1/0 AWG conductor or larger to be installed in parallel, four 1/0 AWG copper conductors must be used instead as the parallel neutral conductors.

In total circular mils, the four 1/0 AWG parallel neutral conductors will exceed the required 3/0 AWG grounding electrode conductor, when considered. The size of the grounding electrode conductor is based on the equivalent area of the parallel service conductors [2400 kcmil (600 kcmil x 4)] per Table 250.66 which is over 1100 kcmil copper.

NEUTRAL CONDUCTOR(S) $\underline{\text{4-1/0 AWG copper conductors}}$

1200SF units (12) - 208/120V 3-phase (1200SF units load calculation covers pages 97 – 103)

1. General Lighting and Receptacle Loads <NEC 220.12, Table 220.12, 220.14(J) and 220.42> (Open porches, garages, unused or unfinished spaces not adaptable for future use not included.)

$$1200SF \times 12 \times 3VA = 43{,}200VA$$

2. Small-Appliance Circuit Load (Portable) <NEC 220.52(A)> (Minimum of two circuits required.)

$$1500VA \times 2 \times 12 = 36{,}000VA$$

3. Laundry Circuit Load <NEC 220.52(B)>

$$1500VA \times 12 = 18{,}000VA$$

TOTAL (Lines 1. - 3.) = 97,200VA
(If Total VA is less than or equal to 120,000VA, step c. is not required)

APPLY DEMAND FACTORS <NEC and Table 220.42>

a. First 3000VA of above TOTAL (At 100%)   =   3000VA

b. 94,200V   x .35   =   32,970VA
    (Total VA – 3001VA up to 117,000VA)

c.   0   x .25   =   0
   (Remainder of TOTAL exceeding 120,000VA)*

       TOTAL (Lines a. - c.)   =   35,970VA

GENERAL LIGHTING and RECEPTACLE, SMALL-APPLIANCE and LAUNDRY LOADS

1. - 3.   LINE LOAD   NEUTRAL LOAD

      35,970VA   35,970VA

4. Appliance Loads (Fastened-In-Place) <NEC 220.53> (Use nameplate rating of each appliance. Electric ranges, dryers, space-heating equipment or air-conditioning equipment not included.)

120V Appliances	VA	No. of Appliances	Total VA
1. Dishwasher	1250	12	15,000
2. Disposal	900	12	10,800
3. Trash Compactor	780	12	9360

4. Microwave	1600	12	19,200
5. Heat-Vent-Light	1600	12	<u>19,200</u>
		(Total 120V Appliances)	73,560

208V Appliances	**VA**	**No. Appliances**	**Total VA**
1. Water Heater	3380	12	<u>40,560</u>
		(Total 208V Appliances)	40,560

APPLIANCES TOTAL = 114,120VA

APPLY DEMAND FACTOR (Applicable, when number of above appliances exceeds four (4) or more.)

(Appliances Total) <u>114,120VA</u> x .75 = 85,590VA

APPLIANCE LOADS

4. <u>LINE LOAD</u> <u>NEUTRAL LOAD (Refer to condition)</u>

85,590VA 55,170VA

Condition
Because the water heaters are 208 volts line-to-line loads they will not contribute to the neutral load as the other appliances and therefore must not be included. As a result, the NEUTRAL LOAD is equal to

114,120VA – 40,560VA x .75 = 55,170VA

5. Clothes Dryers (208/120V-1ϕ) <NEC and Table 220.54> (Use 5000[VA] or nameplate rating, whichever is larger.)

Dryer Load determined per single-phase dryers supplied by 3-phase 4W Feeder or Service

Number of dryers = 12

$\dfrac{12}{3}$ = 4 (number of dryers per phase)

4 x 2 = 8 (number of dryers)

Based on 8 dryers, the demand factor (%) = 60(.60)

5000VA x 8 x .60 = 24,000VA (Demand Load for 2 phases)

$\dfrac{24,000VA}{2\ (phases)}$ = 12,000VA [Demand Load per phase (1ϕ)]

12,000VA [Demand Load per phase (1ϕ)] x 3 = 36,000VA [3-phase (3ϕ) Demand Load]

CLOTHES DRYERS

5. <u>LINE LOAD</u>   <u>NEUTRAL LOAD [NEC 220.61(C)(1)]</u>

    36,000VA   36,000VA

6. Electric Ranges and Other Cooking Appliances <NEC and Table 220.55>

Again, because NEC 220.55 only mentions <u>single-phase ranges</u> being supplied by a 3-phase, 4-wire feeder or service, the described method cannot be used in this situation based upon the exclusive use of non-range cooking appliances that consist of cooktops and wall-mounted ovens. Therefore, this question will only apply the provisions of Columns B (cooking appliances rated between 3½kW to 8¾kW) and C (cooking appliances not rated over 12kW) to derive the service demand load.

Similar to question No. 70, the appliances total kW rating must first be determined

$$
\begin{array}{llll}
\text{Cooktops -} & 12 \text{ x } 6300\text{W} = & 75,600\text{W} \\
\text{WMOs -} & \underline{12} \text{ x } 5400\text{W} = & \underline{64,800\text{W}} \\
& 24 & 140,400\text{W (140.4kW)}
\end{array}
$$

Per Column B

Based upon 24 cooking appliances, the demand factor per Column B is 26 percent (.26) and when applied to the total kW rating yields the following demand load:

$$140.4\text{kW x } .26 = 36.5\text{kW (36,500W) (rounded off)}$$

Per Column C

Determine the maximum demand of the cooking appliances and compare with the calculated value per Column B. Use the lowest value.

<u>Maximum Demand for 24 Cooking Appliances</u>

Maximum Demand = 39kW (39,000W)

Applying the lowest value, the service demand for the cooking appliances is 36.5kW.

Similar to the single and three-phase demand loads calculated based upon the provisions provided in NEC 220.55, the load of the 24 cooking appliances about the 208/120V, 4-wire, 3-phase system supplying the twelve (12) unit apartment complex can be evenly distributed.

Therefore, between each two-phases of this 208/120V three-phase system, four cooktops and wall-mounted ovens can be evenly placed and balanced about the system, resulting to a demand load of 12.17kW (36.5kW / 3) per two-phase connections.

If these cooking appliances were <u>single-phase ranges</u> instead and calculated according to NEC 220.55, the LINE and NEUTRAL LOADS would have yielded higher values if determined based upon the provisions of Columns B and C.

RANGE/COOKING APPLIANCES

6. <u>LINE LOAD</u>   <u>NEUTRAL LOAD [NEC 220.61(C)(1)]</u>
                                          (Refer to condition)

    36,500VA   20,740VA

<u>Condition</u>
Because the cooktops are 208 volts line-to-line loads they will not contribute to the neutral load and therefore must not be included. As a result, the RANGE/COOKING APPLIANCES - LINE LOAD must be recalculated omitting the cooktops to determine the NEUTRAL LOAD.

NEUTRAL LOAD – 208/120V (Wall-mounted Ovens [WMOs])

Number of appliances (with neutral connections) = 12

The appliances total kW rating.

WMOs -   <u>12</u> x 5400W =   <u>64,800W (64.8kW)</u>

Per Column B

Based upon 12 cooking appliances, the demand factor per Column B is 32 percent (.32) and when applied to the total kW rating yields the following demand load:

64.8kW x .32 = 20.74kW (20,740W) (rounded off)

Per Column C

Determine the maximum demand of the cooking appliances and compare with the calculated value per Column B. Use the lowest value.

<u>Maximum Demand for 12 Cooking Appliances</u>

Maximum Demand = 27kW (27,000W)

Compared to the Maximum Demand of Column C (27,000VA) - Use 20,740VA as the neutral load contributed by the wall-mounted ovens.

7. Heating and Air-Conditioning (AC) Equipment <NEC References - 220.50, 220.51, 220.60, Table 430. 248 and 440.6(A)> (Include VA rating of air handler [blower].)

Heating Load

Electric Heat

$$15kW (15,000VA) \times \underline{12} = 180,000VA$$

AC Load

$$5472VA \times \underline{12} = 65,664VA$$

Heating load is larger.

Electric Heat   Blowers

Line Load = $\underline{180,000VA}$ + $\underline{1320VA} \times \underline{12}$ = $\underline{195,840VA}$

HEATING and AC

7.	LINE LOAD	NEUTRAL LOAD
	195,840VA	0

8. Largest Motor <NEC 430.17, 430.24, 440.7 and 440.33> (Use motor with highest full-load current [FLC] regardless of voltage rating.)

Largest Motor (LM) = 900VA (7.5A) disposal rated for 120 volts

$\underline{900VA} \times .25 = \underline{225VA}$
(LM)

LARGEST MOTOR

8.	LINE LOAD	NEUTRAL LOAD (120V motors only)
	225VA	225VA

TOTAL DEMAND LOAD (LINE and NEUTRAL) (Add lines 1. – 8.)

	LINE LOAD	NEUTRAL LOAD
1. - 3.	35,970VA	35,970VA
4.	85,590VA	55,170VA
5.	36,000VA	36,000VA
6.	36,500VA	20,740VA

7.	195,840VA		0
8.	**225VA**		**225VA**
	392,940VA		111,141VA

DWELLING'S OPERATING LINE VOLTAGE - 208V(3ϕ)
(Given operating voltage or as determined per test examination)

CALCULATE MINIMUM LINE and NEUTRAL LOADS
(Divide Total Demand Load **[VA]** by operating line voltage **[V]**)

LINE LOAD = 392,940VA / 208V x 1.732 = 1090.72A

NEUTRAL LOAD

Because the DRYER (12,000VA) and RANGE (7776VA) loads are fed from circuits which are prohibited per NEC 260.61(C)(1), the reduction of both neutral loads per NEC 260.61(B) cannot be applied. Combined DRYER and RANGE neutral loads = 19,776VA.

Permitted Reduction

111,141VA – 19,776VA = 91,365VA (Permitted)

91,365VA / 208V x 1.732 = ~~253.61A~~* 237.52A

*Where the neutral load exceeds 200A, NEC 220.61(B)(2) permits the load to be reduced by 70 percent. Complete the following to determine permitted Neutral Demand Load.

(1) 253.61 ___ A – 200A = 53.61 ___ A x .70 = 37.53 ___ A
    (Neutral Load)*   (Remainder)   (Permitted Reduction)

(2) 37.53 ___ A + 200A = 237.53 ___ A
    (Permitted Reduction)   (Permitted Neutral Demand Load)

Prohibited Reduction

19,776VA / 208V x 1.732 = 54.89A

NEUTRAL LOAD = 237.53A + 54.89A = 292.42A

SIZE SERVICE (Size of service based on the calculated **LINE LOAD**)

SIZE SERVICE REQUIRED (minimum) 1200A
(Single rating or combination - Main overcurrent device(s) to total **1200A**)

The conductors or combinations selected for this installation are only limited for this question. Conductor ampacity derating is not applied in sizing the service and neutral conductors.

SIZING FEEDER/SERVICE CONDUCTORS

Using NEC 310.10(H)(1) and Table 310.15(B)(16), as a minimum based on NEC 240.4(C), use three 600 kcmil copper conductors (75°C) per phase which have an individual ampacity of 420A, as the service conductors.

FEEDER/SERVICE CONDUCTORS 3-600 kcmil copper conductors per line

SIZING NEUTRAL CONDUCTOR(S)

Based on the calculated neutral load and the use of parallel service conductors, the neutral conductors must also be installed in parallel per NEC 310.10(H)(1). Because the service conductors will consist of three parallel conductors per phase, the neutral conductors must also consist of three parallel conductors based on the calculated neutral load being divided by three which results to 131.68A (395.03A / 3). Per Table 310.15(B)(16), at 75°C, a 1/0 AWG copper conductor which has a rated ampacity of 150A will satisfy the calculated neutral load being distributed amongst three such conductors. Unlike previous requirements for the other units, the sizing of the parallel neutral conductors for this unit is derived directly from the initial calculation which requires the use of 1/0 AWG copper conductors per NEC 310.10(H)(1).

In total circular mils, the three 1/0 AWG parallel neutral conductors will exceed the required 3/0 AWG grounding electrode conductor, when considered. The size of the grounding electrode conductor is based on the equivalent area of the parallel service conductors [1800 kcmil (600 kcmil x 3)] per Table 250.66 which is over 1100 kcmil copper.

NEUTRAL CONDUCTOR(S) 3-1/0 AWG copper conductors

1400SF units (10) - 208/120V 3-phase (1400SF units load calculation covers pages 103 – 109)

1. General Lighting and Receptacle Loads <NEC 220.12, Table 220.12, 220.14(J) and 220.42> (Open porches, garages, unused or unfinished spaces not adaptable for future use not included.)

$$1400SF \times 10 \times 3VA = 42,000VA$$

2. Small-Appliance Circuit Load (Portable) <NEC 220.52(A)> (Minimum of two circuits required)

$$1500VA \times 2 \times 10 = 30,000VA$$

3. Laundry Circuit Load <NEC 220.52(B)>

$$1500VA \times 10 = 15,000VA$$

$$TOTAL \text{ (Lines 1. - 3.)} = 87,000VA$$
(If Total VA is less than or equal to 120,000VA, step c. is not required)

APPLY DEMAND FACTORS <NEC and Table 220.42>

a. First 3000VA of above TOTAL (At 100%)   =   3000VA

b. 84,000VA   x .35   = 29,400VA
   (Total VA – 3001VA up to 117,000VA)

c.   0   =   0
   (Remainder of TOTAL exceeding 120,000VA)*

            TOTAL (Lines a. - c.)   = 32,400VA

GENERAL LIGHTING and RECEPTACLE, SMALL-APPLIANCE and LAUNDRY LOADS

1. - 3. <u>LINE LOAD</u>   <u>NEUTRAL LOAD</u>

    32,400VA   32,400VA

4.   Appliance Loads (Fastened-In-Place) <NEC 220.53> (Use nameplate rating of each appliance. Electric ranges, dryers, space-heating equipment or air-conditioning equipment not included.)

120V Appliances	VA	No. of Appliances	Total VA
1. Dishwasher	1250	10	12,500
2. Disposal	900	10	9000
3. Trash Compactor	780	10	7800
4. Microwave	1800	10	18,000
5. Heat-Vent-Light	1750	10	17,500
		(Total 120V Appliances)	64,800

208V Appliances	VA	No. Appliances	Total VA
1. Water Heater	3380	10	33,800
		(Total 208V Appliances)	33,800

            APPLIANCES TOTAL   =   98,600VA

APPLY DEMAND FACTOR (Applicable, when number of above appliances exceeds four (4) or more.)

        (Appliances Total) <u>98,600VA</u> x .75 = 73,950VA

APPLIANCE LOADS

4.   <u>LINE LOAD</u>   <u>NEUTRAL LOAD</u> (Refer to condition)

    73,950VA   48,600VA

<u>Condition</u>
Because the water heaters are 208 volts line-to-line loads they will not contribute to the neutral load as the other appliances and therefore must not be included. As a result, the NEUTRAL LOAD is equal to

$$98,600VA - 33,800VA \text{ x } .75 = 48,600VA$$

5. Clothes Dryers (208/120V-1φ) <NEC and Table 220.54> (Use 5000W [VA] or nameplate rating, whichever is larger.)

Dryer Load determined per single-phase dryers supplied by 3-phase 4W Feeder or Service

Number of dryers = 10 $\quad \dfrac{10}{3} = 3.3 \text{ 4 (number of dryers per phase)}$

$4 \text{ x } 2 = 8$ (number of dryers)

Based on 8 dryers,

the demand factor (%) = 60(.60)

$5000VA \text{ x } 8 \text{ x } .60 = 24,000VA$ (Demand Load for 2 phases)

$\dfrac{24,000VA}{2 \text{ (phases)}} = 12,000VA$ [Demand Load per phase (1φ)]

12,000VA [Demand Load per phase (1φ)] x 3 = 36,000VA [3-phase (3φ) Demand Load]

CLOTHES DRYERS

5.	LINE LOAD	NEUTRAL LOAD [NEC 220.61(C)(1)]
	36,000VA	36,000VA

6. Electric Ranges and Other Cooking Appliances <NEC and Table 220.55>

Again, because NEC 220.55 only mentions <u>single-phase ranges</u> being supplied by a 3-phase, 4-wire feeder or service, the described method cannot be used in this situation based upon the exclusive use of non-range cooking appliances that consist of cooktops and wall-mounted ovens. Therefore, this question will only apply the provisions of Columns B and C to derive the service demand load.

Cooktops -	10 x 6700W =	67,000W
WMOs -	$\dfrac{10 \text{ x } 5800W}{20}$ =	$\dfrac{58,000W}{125,000W \text{ (125kW)}}$

Per Column B

Based upon 20 cooking appliances, the demand factor per Column B is 28 percent (.28) and when applied to the total kW rating yields the following demand load:

$$125kW \times .28 = 35kW \ (35,000W)$$

Per Column C

Determine the maximum demand of the cooking appliances and compare with the calculated value per Column B. Use the lowest value.

<u>Maximum Demand for 20 Cooking Appliances</u>

Maximum Demand = 35kW (35,000W)

The service demand for the cooking appliances is 35kW based upon the provisions of either Column B or C.

Similar to the single and three-phase demand loads calculated based upon the provisions provided in NEC 220.55, the load of the 20 cooking appliances about the 208/120V, 4-wire, 3-phase system supplying the ten (10) unit apartment complex cannot be evenly distributed.

Therefore, between each two-phases of this 208/120V three-phase system, four (4) cooktops and wall-mounted ovens can be placed between any one of the three two-phase combinations, while the six (6) remaining cooktops and wall-mounted ovens can be evenly placed between the two remaining two-phase connections.

RANGE/COOKING APPLIANCES

6.	<u>LINE LOAD</u>	<u>NEUTRAL LOAD [NEC 220.61(C)(1)]</u> (Refer to condition)
	35,000VA	19,720VA

Condition
Because the cooktops are 208 volts line-to-line loads they will not contribute to the neutral load and therefore must not be included. As a result, the RANGE/COOKING APPLIANCES - LINE LOAD must be recalculated omitting the cooktops to determine the NEUTRAL LOAD.

NEUTRAL LOAD – 208/120V (Wall-mounted Ovens [WMOs])

Number of appliances (with neutral connections) = 10

The appliances total kW rating.

$$\text{WMOs -} \quad \underline{10} \times 5800W = \quad \underline{58,000W \ (58kW)}$$

Per Column B

Based upon 10 cooking appliances, the demand factor per Column B is 34 percent (.34) and when applied to the total kW rating yields the following demand load:

$$58kW \times .34 = 19.72kW \ (19,720W)$$

Per Column C

Determine the maximum demand of the cooking appliances and compare with the calculated value per Column B. Use the lowest value.

<u>Maximum Demand for 10 Cooking Appliances</u>

$$\text{Maximum Demand} = 25kW \ (25,000W)$$

Compared to the Maximum Demand of Column C (25,000VA) - Use 19,720VA as the neutral load contributed by the wall-mounted ovens.

7. Heating and Air-Conditioning (AC) Equipment <NEC References - 220.50, 220.51, 220.60, Table 430. 248 and 440.6(A)> (Include VA rating of air handler [blower].)

Heating Load

$$18kW \ (18,000VA) \times \underline{10} = 180,000VA$$

AC Load

$$6384VA \times \underline{10} = 63,840VA$$

Heating load is larger.

$$\begin{array}{cc} \text{Electric Heat} & \text{Blowers} \\ \text{Line Load} = \underline{180,000VA} + \underline{1416VA} \times \underline{10} = \underline{194,160VA} \end{array}$$

HEATING and AC

7.	LINE LOAD	NEUTRAL LOAD
	194,160VA	0

8. Largest Motor <NEC 430.17, 430.24, 440.7 and 440.33> (Use motor with highest full-load current [FLC] regardless of voltage rating.)

Largest Motor (LM) = 900VA (7.5A) blower rated for 120 volts

900VA x .25 = 225VA
(LM)

LARGEST MOTOR

8.   LINE LOAD   NEUTRAL LOAD (120V motors only)

      225VA   225VA

TOTAL DEMAND LOAD (LINE and NEUTRAL) (Add lines 1. – 8.)

	LINE LOAD	NEUTRAL LOAD
1. - 3.	32,400VA	32,400VA
4.	73,950VA	48,600VA
5.	36,000VA	12,000VA
6.	42,000VA	8350VA
7.	194,160VA	0
8.	225VA	225VA
	378,735VA	101,575VA

DWELLING'S OPERATING LINE VOLTAGE - 208V(3φ)
(Given operating voltage or as determined per test examination)

CALCULATE MINIMUM LINE and NEUTRAL LOADS
(Divide Total Demand Load **[VA]** by operating line voltage **[V]**)

LINE LOAD = 378,735VA / 208V x 1.732 = 1051.29A

NEUTRAL LOAD
Because the DRYER (12,000VA) and RANGE (8350VA) loads are fed from circuits which are prohibited per NEC 260.61(C)(1), the reduction of both neutral loads per NEC 260.61(B) cannot be applied. Combined DRYER and RANGE neutral loads = 20,350VA.

Permitted Reduction

      101,575VA – 20,350VA = 81,225VA (Permitted)

      81,225VA / 208V x 1.732 = ~~225.46A~~* 217.82A

*Where the neutral load exceeds 200A, NEC 220.61(B)(2) permits the load to be reduced by 70 percent. Complete the following to determine permitted Neutral Demand Load.

(1) 225.46   A – 200A = 25.46   A x .70 = 17.82A
    (Neutral Load)*   (Remainder)   (Permitted Reduction)

(2) <u>17.82</u>   A + 200A = <u>217.82</u>   A
   (Permitted Reduction)   (Permitted Neutral Demand Load)

Prohibited Reduction

<u>20,350</u>VA / <u>208V</u> x 1.732 = <u>56.49A</u>

NEUTRAL LOAD = <u>217.82</u>A + <u>56.49</u>A = <u>274.31</u>A

SIZE SERVICE (Size of service based on the calculated **LINE LOAD**)

SIZE SERVICE REQUIRED (minimum) <u>1200A</u>
(Single rating or combination - Main overcurrent device(s) to total **1200A**)

The conductors or combinations selected for this installation are only limited for this question. Conductor ampacity derating is not applied in sizing the service and neutral conductors.

SIZING FEEDER/SERVICE CONDUCTORS

Using NEC 310.10(H)(1) and Table 310.15(B)(16), as a minimum based on NEC 240.4(C), use three 600 kcmil copper conductors (75°C) per phase which have an individual ampacity of 420A, as the service conductors.

FEEDER/SERVICE CONDUCTORS <u>3-600 kcmil copper conductors per line</u>

SIZING NEUTRAL CONDUCTOR(S)

Based on the calculated neutral load and the use of parallel service conductors, the neutral conductors must also be installed in parallel per NEC 310.10(H)(1). Because the service conductors will consist of three parallel conductors per phase, the neutral conductors must also consist of three parallel conductors based on the calculated neutral load being divided by three which results to 91.44A (274.31A / 3). Per Table 310.15(B)(16), at 75°C, a 3 AWG copper conductor which has a rated ampacity of 100A will satisfy the calculated neutral load being distributed amongst three such conductors. However, because NEC 310.10(H)(1) only allows a 1/0 AWG conductor or larger to be installed in parallel, three 1/0 AWG copper conductors must be used instead as the parallel neutral conductors.

In total circular mils, the three 1/0 AWG parallel neutral conductors will exceed the required 3/0 AWG grounding electrode conductor, when considered. The size of the grounding electrode conductor is based on the equivalent area of the parallel service conductors [1800 kcmil (600 kcmil x 3)] per Table 250.66 which is over 1100 kcmil copper.

NEUTRAL CONDUCTOR(S) <u>3-1/0 AWG copper conductors</u>

SINGLE-PHASE LOAD CALCULATIONS PER INDIVIDUAL UNITS AT 208/120V

For "Multifamily Dwelling" reference format see **Worksheet A** (Volume 4) – STANDARD LOAD CALCULATIONS FOR ONE-FAMILY DWELLING.

650SF unit - 208/120V 1-phase (650SF unit load calculation covers pages 110 – 113)

1. General Lighting and Receptacle Loads <NEC 220.12, Table 220.12, 220.14(J) and 220.42> (Open porches, garages, unused or unfinished spaces not adaptable for future use not included.)

$$650SF \text{ x } 3VA = 1950VA$$

2. Small-Appliance Circuit Load (Portable) <NEC 220.52(A)> (Minimum of two circuits required.)

$$1500VA \text{ x } 2 = 3000VA$$

3. Laundry Circuit Load <NEC 220.52(B)[1]>

[1]650SF unit not equipped with washer/dryer connections.

0

TOTAL (Lines 1. - 3.) = 4950VA
(If Total VA is less than or equal to 120,000VA, step c. is not required)

APPLY DEMAND FACTORS (NEC and Table 220.42)

a. First 3000VA of above TOTAL (At 100%) = 3000.0VA

b. 1950VA   x .35   =   682.5VA
   (Total VA – 3001VA up to 117,000VA)

c.   0   x .25   =   0
   (Remainder of TOTAL exceeding 120,000VA)*

TOTAL (Lines a. - c.)   = 3682.5VA

GENERAL LIGHTING and RECEPTACLE, SMALL-APPLIANCE and LAUNDRY LOADS

1. - 3. <u>LINE LOAD</u>   <u>NEUTRAL LOAD</u>

3682.5VA   3682.5VA

4. Appliance Loads (Fastened-In-Place) <NEC 220.53> (Use nameplate rating of each appliance. Electric ranges, dryers, space-heating equipment or air-conditioning equipment not included.)

120V Appliances	**VA Rating**
1. Dishwasher	900
2. Disposal	600
3. Microwave	1000
4. Heat-Vent-Light	1300
	3800 (Total 120V Appliances)

208V Appliances	**VA Rating**
1. Water Heater	3380 (Total 208V Appliance)
	3380

APPLIANCES TOTAL   =   7180VA

APPLY DEMAND FACTOR (Applicable, when number of above appliances exceeds four (4) or more.)

(Appliances Total) 7180VA x .75 =   5385VA

APPLIANCE LOADS

4.   LINE LOAD   NEUTRAL LOAD (Refer to condition)

5385VA   2850VA

Condition
Because the water heater is a 208 volts line-to-line load it will not contribute to the neutral load as the other appliances and therefore must not be included. As a result, the NEUTRAL LOAD is equal to

7180VA - 3380VA x .75 = 2850VA

5. Clothes Dryer[2] (208/120V-1φ) <NEC and Table 220.54> (Use 5000W [VA] or nameplate rating, whichever is larger.)

[2] 650SF unit not equipped with washer/dryer connections.

CLOTHES DRYER

5.   LINE LOAD   NEUTRAL LOAD

0   0

6. Electric Ranges and Other Cooking Appliances (208/120V-1φ) <NEC and Table 220.55>

Minimum Demand Load determined per Columns B or C to Table 220.55

Applying Column B
Demand Percent (1 appliance) = .80
(6.1kW) 6100W x .80 = 4880W

Applying Column C
Maximum Demand (1 appliance)
8000W

Use smallest demand load of Columns B (4880W) and C (8000W)

RANGE/COOKING APPLIANCES

6.	LINE LOAD	NEUTRAL LOAD [NEC 220.61(C)(1)]
	4880VA	4880VA

7. Heating and Air-Conditioning Equipment <NEC 220.50, 220.51, 220.60, Table 430.248 and 440.6(A)> (Include VA rating of air handler [blower].)

Heating Load (larger)

10kW (10,000VA)

AC Load

3648VA

$$\text{Line Load} = \underset{\text{Electric Heat}}{\underline{10{,}000\text{VA}}} + \underset{\text{Blower}}{\underline{1044\text{VA}}} = \underline{11{,}044\text{VA}}$$

HEATING and AC

7.	LINE LOAD	NEUTRAL LOAD
	11,044VA	0

8. Largest Motor <NEC 430.17, 430.24, 440.7 and 440.33> (Use motor with highest full-load current [FLC] regardless of voltage rating.)

Largest Motor (LM) = 1044VA (5.02A) blower rated for 208 volts

$$\underset{\text{(LM)}}{\underline{1044\text{VA} \times .25 = 261\text{VA}}}$$

LARGEST MOTOR

8.   LINE LOAD   NEUTRAL LOAD (120V motors only)

         261VA   0

TOTAL DEMAND LOAD (LINE and NEUTRAL) (Add lines 1. − 8.)

	LINE LOAD	NEUTRAL LOAD
1. - 3.	3682.5VA	3682.5VA
4.	5385.0VA	2850.0VA
5.	0	0
6.	4880.0VA	4880.0VA
7.	11,044.0VA	0
8.	261.0VA	0
	25,252.5VA	11,332.5VA

DWELLING'S OPERATING LINE VOLTAGE - 208V
(Given operating voltage or as determined per test examination)

CALCULATE MINIMUM LINE and NEUTRAL LOADS
(Divide Total Demand Load [VA] by operating line voltage [V])

LINE LOAD = 25,252.5VA / 208V = 121.41A
NEUTRAL LOAD = 11,332.5VA / 208V = 54.48A

SIZE SERVICE (Size of service based on the calculated **LINE LOAD**)

SIZE SERVICE REQUIRED (minimum) 125A
(Single rating - Main overcurrent device **125A**)

SIZING FEEDER/SERVICE CONDUCTORS

Applying NEC 240.4(B) and Table 310.15(B)(16), as a minimum use 2 AWG copper conductors (75°C) which have an ampacity of 115A as the service conductors.

FEEDER/SERVICE CONDUCTORS 2 AWG copper conductors

SIZING NEUTRAL CONDUCTOR(S)

Using Table 310.15(B)(16), as a minimum, use a 6 AWG copper conductor (75°C) which has an ampacity of 65A, as the neutral conductor. The use of a 6 AWG copper conductor as the neutral is in compliance with NEC 250.24(C)(1) and Table 250.66.

NEUTRAL CONDUCTOR(S) 6 AWG copper conductor

900SF unit - 208/120V 1-phase (900SF unit load calculation covers pages 114 – 117)

1. General Lighting and Receptacle Loads <NEC 220.12, Table 220.12, 220.14(J) and 220.42> (Open porches, garages, unused or unfinished spaces not adaptable for future use not included.)

$$900SF \times 3VA = 2700VA$$

2. Small-Appliance Circuit Load (Portable) <NEC 220.52(A)> (Minimum of two circuits required.)

$$1500VA \times 2 = 3000VA$$

3. Laundry Circuit Load <NEC 220.52(B)>

$$1500VA \times 1 = 1500VA$$

TOTAL (Lines 1. - 3.) = 7200VA
(If Total VA is less than or equal to 120,000VA, step c. is not required)

APPLY DEMAND FACTORS <NEC and Table 220.42>

a. First 3000VA of above TOTAL (At 100%)   = 3000VA

b. 4200VA   x .35   = 1470VA
   (Total VA – 3001VA up to 117,000VA)

c.   0   x .25   =   0
   (Remainder of TOTAL exceeding 120,000VA)*

TOTAL (Lines a. - c.)   = 4470VA

GENERAL LIGHTING and RECEPTACLE, SMALL-APPLIANCE and LAUNDRY LOADS

1. - 3. LINE LOAD	NEUTRAL LOAD
4470VA	4470VA

4. Appliance Loads (Fastened-In-Place) <NEC 220.53> (Use nameplate rating of each appliance. Electric ranges, dryers, space-heating equipment or air-conditioning equipment not included.)

120V Appliances	**VA Rating**
1. Dishwasher	900
2. Disposal	600

3. Microwave   1200
4. Heat-Vent-Light   1600

                                              4300 (Total 120V Appliances)

208V Appliances   **VA Rating**
1. Water Heater   3380 (Total 208V Appliance)

                                              3380

               APPLIANCES TOTAL   =   7680VA

APPLY DEMAND FACTOR (Applicable, when number of above appliances exceeds four (4) or more.)

               (Appliances Total) 7680VA x .75 =   5760VA

APPLIANCE LOADS

4.   **LINE LOAD**   **NEUTRAL LOAD (Refer to condition)**

     5760VA   3225VA

Condition
Because the water heater is a 208 volts line-to-line load it will not contribute to the neutral load as the other appliances and therefore must not be included. As a result, the NEUTRAL LOAD is equal to

                    7680VA - 3380VA x .75 = 3225VA

5. Clothes Dryer (208/120V-1φ) <NEC and Table 220.54> (Use 5000W [VA] or nameplate rating, whichever is larger.)

CLOTHES DRYER

5.   **LINE LOAD**   **NEUTRAL LOAD [NEC 220.61(C)(1)]**

     5000VA   5000VA

6. Electric Ranges and Other Cooking Appliances (208/120V-1φ) <NEC and Table 220.55>

Minimum Demand Load determined per Columns B or C to Table 220.55.

Applying Column B
Demand Percent (1 appliance) = .80
(7.8kW) 7800W x .80 = 6240W

Applying Column C
Maximum Demand (1 appliance)
8000W

Use smallest demand load of Columns B (6240W) and Column C (8000W)

RANGE/COOKING APPLIANCES

6.   <u>LINE LOAD</u>   <u>NEUTRAL LOAD [NEC 220.61(C)(1)]</u>

6240VA   6240VA

7. Heating and Air-Conditioning Equipment <NEC 220.50, 220.51, 220.60, Table 430.248 and 440.6(A)> (Include VA rating of air handler [blower].)

Heating Load

12kW (12,000VA)

AC Load

4560VA

Heating load is larger.

Line Load = $\underset{\text{Electric Heat}}{\underline{12,000VA}}$ + $\underset{\text{Blower}}{\underline{1176VA}}$ = $\underline{13,176V}$

HEATING and AC

7.   <u>LINE LOAD</u>   <u>NEUTRAL LOAD</u>

13,176VA   0

8. Largest Motor <NEC 430.17, 430.24, 440.7 and 440.33> (Use motor with highest full-load current [FLC] regardless of voltage rating.)

Largest Motor (LM) = 1176VA (5.65A) blower rated for 208 volts

$\underset{\text{(LM)}}{\underline{1176VA}}$ x .25 = $\underline{294VA}$

LARGEST MOTOR

8.	LINE LOAD	NEUTRAL LOAD (120V motors only)
	294VA	0

TOTAL DEMAND LOAD (LINE and NEUTRAL) (Add lines 1. – 8.)

	LINE LOAD	NEUTRAL LOAD
1. - 3.	4470VA	4470VA
4.	5760VA	3225VA
5.	5000VA	5000VA
6.	6240VA	6240VA
7.	13,176VA	0
8.	294VA	0
	34,940VA	18,935VA

DWELLING'S OPERATING LINE VOLTAGE - 208V
(Given operating voltage or as determined per test examination)

CALCULATE MINIMUM LINE and NEUTRAL LOADS
(Divide Total Demand Load [VA] by operating line voltage [V])

LINE LOAD = 34,940VA / 208V = 167.98A
NEUTRAL LOAD = 18,935VA / 208V = 91.03A

SIZE SERVICE (Size of service based on the calculated **LINE LOAD**)

SIZE SERVICE REQUIRED (minimum) 175A
(Single rating - Main overcurrent device175A)

SIZING FEEDER/SERVICE CONDUCTORS

Using Table 310.15(B)(16), as a minimum use 2/0 AWG copper conductors (75°C) which has an ampacity of 175A as the service conductors.

FEEDER/SERVICE CONDUCTORS 2/0 AWG copper conductors

SIZING NEUTRAL CONDUCTOR(S)

Using Table 310.15(B)(16), as a minimum, use a 3 AWG copper conductor (75°C) which has an ampacity of 100A, as the neutral conductor. The use of a 3 AWG copper conductor as the neutral is in compliance with NEC 250.24(C)(1) and Table 250.66.

NEUTRAL CONDUCTOR(S) 3 AWG copper conductor

1200SF unit - 208/120V 1-phase (1200SF unit load calculation covers pages 128 – 132)

1. General Lighting and Receptacle Loads <NEC 220.12, Table 220.12, 220.14(J) and 220.42> (Open porches, garages, unused or unfinished spaces not adaptable for future use not included.)

$$1200SF \times 3VA = 3600VA$$

2. Small-Appliance Circuit Load (Portable) <NEC 220.52(A)> (Minimum of two circuits required.)

$$1500VA \times 2 = 3000VA$$

3. Laundry Circuit Load <NEC 220.52(B)>

$$1500VA \times 1 = 1500VA$$

TOTAL (Lines 1. - 3.) = 8100VA
(If Total VA is less than or equal to 120,000VA, step c. is not required)

APPLY DEMAND FACTORS <NEC and Table 220.42>

a. First 3000VA of above TOTAL (At 100%) = 3000VA

b. 5100VA   x .35   =   1785VA
   (Total VA – 3001VA up to 117,000VA)

c.   0   x .25   =   0
   (Remainder of TOTAL exceeding 120,000VA)*

     TOTAL (Lines a. - c.)   =   4785VA

GENERAL LIGHTING and RECEPTACLE, SMALL-APPLIANCE and LAUNDRY LOADS

1. - 3. LINE LOAD	NEUTRAL LOAD
4785VA	4785VA

4. Appliance Loads (Fastened-In-Place) <NEC 220.53> (Use nameplate rating of each appliance. Electric ranges, dryers, space-heating equipment or air conditioning equipment not included.)

120V Appliances	**VA Rating**
1. Dishwasher	1250
2. Disposal	900
3. Trash Compactor	780

4. Microwave	1600
5. Heat-Vent-Light	<u>1600</u>
	6130 (Total 120V Appliances)

208V Appliances	**VA Rating**
1. Water Heater	<u>3380</u> (Total 208V Appliance)
	3380

APPLIANCES TOTAL   =   9510VA

APPLY DEMAND FACTOR (Applicable, when number of above appliances exceeds four (4) or more.)

(Appliances Total) <u>9510VA</u> x .75 = 7132.5VA

APPLIANCE LOADS

4.	LINE LOAD	NEUTRAL LOAD (Refer to condition)
	7132.5VA	4597.5VA

Condition

Because the water heater is a 208 volts line-to-line load it will not contribute to the neutral load as the other appliances and therefore must not be included. As a result, the NEUTRAL LOAD is equal to

9510VA - 3380VA x .75 = 4597.5VA

5. Clothes Dryer (208/120V-1ϕ) <NEC and Table 220.54> (Use 5000W [VA] or nameplate rating, whichever is larger.)

CLOTHES DRYER

5.	LINE LOAD	NEUTRAL LOAD [NEC 220.61(C)(1)]
	5000VA	5000VA

6. Electric Ranges and Other Cooking Appliances (208/120V-1ϕ) <NEC and Table 220.55>

Minimum Demand Load determined per Columns B or C to Table 220.55.

Applying Column B

Demand Percent (2 appliances) = .65
(5.4kW) 5400W + (6.3kW) 6300W x .65 = 7605W

Applying Column C

Maximum Demand (2 appliances)
11kW(11,000W)

Use the smallest demand load of Column B (7605W) Column C (11000W).

REMINDER: Although it may appear that the initial provision of **Note 4** to Table 220.55 could have been applied toward the use of the two cooking appliances, it cannot. Because both cooking appliances are used toward calculating the service load for this occupancy such provision is not applicable. The provisions of **Note 4** to Table 220.55 are only applicable for branch-circuit loads and not for service and feeder load calculations.

NEUTRAL LOAD

Based on the exclusive use of the WMO per Column B the neutral load is,

$$(5.4kW) 5400W \times .80 = 4320VA$$

RANGE/COOKING APPLIANCES

6.   <u>LINE LOAD</u>   <u>NEUTRAL LOAD [NEC 220.61(C)(1)]</u>

7605VA   4320VA

7. Heating and Air-Conditioning Equipment <NEC 220.50, 220.51, 220.60, Table 430.248 and 440.6(A)> (Include VA rating of air handler [blower].)

Heating Load

15kW (15,000VA)

AC Load

5472VA

Heating load is larger

Electric Heat   Blower
Line Load = <u>15,000VA</u> + <u>1320VA</u> = <u>16,320VA</u>

HEATING and AC

7.   <u>LINE LOAD</u>   <u>NEUTRAL LOAD</u>

16,320VA   0

8. Largest Motor <NEC 430.17, 430.24, 440.7 and 440.33> (Use motor with highest full-load current [FLC] regardless of voltage rating.)

Largest Motor (LM) = 900VA (7.5A) blower rated for 120 volts

$\underline{900VA}$ x .25 = $\underline{225VA}$
(LM)

LARGEST MOTOR

8.	LINE LOAD	NEUTRAL LOAD (120V motors only)
	225VA	225VA

TOTAL DEMAND LOAD (LINE and NEUTRAL) (Add lines 1. – 8.)

	LINE LOAD	NEUTRAL LOAD
1. - 3.	4785.0VA	4785.0VA
4.	7132.5VA	4597.5VA
5.	5000.0VA	5000.0VA
6.	7605.0VA	4320.0VA
7.	16,320.0VA	0
8.	225.0VA	225.0VA
	41,067.5 VA	18,927.5VA

DWELLING'S OPERATING LINE VOLTAGE - $\underline{208}$V
(Given operating voltage or as determined per test examination)

CALCULATE MINIMUM LINE and **NEUTRAL LOADS**
(Divide Total Demand Load [VA] by operating line voltage [V])

LINE LOAD = $\underline{41,067.5}$VA / $\underline{208}$V = $\underline{197.44}$A
NEUTRAL LOAD = $\underline{18,927.5}$VA / $\underline{208}$V = $\underline{90.99}$A

SIZE SERVICE (Size of service based on the calculated **LINE LOAD**)

SIZE SERVICE REQUIRED (minimum) $\underline{200A}$
(Single rating - Main overcurrent device **200A**)

SIZING FEEDER/SERVICE CONDUCTORS

Using Table 310.15(B)(16), as a minimum use 3/0 AWG copper conductors (75°C) which has an ampacity of 200A as the service conductors.

FEEDER/SERVICE CONDUCTORS $\underline{3/0\ AWG\ copper\ conductors}$

SIZING NEUTRAL CONDUCTOR(S)

Using Table 310.15(B)(16), as a minimum, use a 3 AWG copper conductor (75°C) which has an ampacity of 100A, as the neutral conductor. The use of a 3 AWG copper conductor as the neutral is in compliance with NEC 250.24(C)(1) and Table 250.66.

NEUTRAL CONDUCTOR(S) 3 AWG copper conductor

1400SF unit - 208/120V 1-phase (1400SF unit load calculation covers pages 132 – 136)

1. General Lighting and Receptacle Loads <NEC 220.12, Table 220.12, 220.14(J) and 220.42> (Open porches, garages, unused or unfinished spaces not adaptable for future use not included.)

$$1400SF \times 3VA = 4200VA$$

2. Small-Appliance Circuit Load (Portable) <NEC 220.52(A)> (Minimum of two circuits required.)

$$1500VA \times 2 = 3000VA$$

3. Laundry Circuit Load <NEC 220.52(B)>

$$1500VA \times 1 = 1500VA$$

$$TOTAL (Lines 1. - 3.) = 8700VA$$
(If Total VA is less than or equal to 120,000VA, step c. is not required)

APPLY DEMAND FACTORS (NEC and Table 220.42)

a. First 3000VA of above TOTAL (At 100%) = 3000VA

b. 5700VA   x .35   = 1995VA
   (Total VA – 3001VA up to 117,000VA)

c.   0   =   0
   (Remainder of TOTAL exceeding 120,000VA)*

TOTAL (Lines a. - c.)   = 4995VA

GENERAL LIGHTING and RECEPTACLE, SMALL-APPLIANCE and LAUNDRY LOADS

1. - 3. LINE LOAD   NEUTRAL LOAD

   4995VA   4995VA

4. Appliance Loads (Fastened-In-Place) <NEC 220.53> (Use nameplate rating of each appliance. Electric ranges, dryers, space-heating equipment or air-conditioning equipment not included.)

120V Appliances	**VA Rating**
1. Dishwasher	1250
2. Disposal	900
3. Trash Compactor	780
4. Microwave	1800
5. Heat-Vent-Light	<u>1750</u>
	6480 (Total 120V Appliances)

208V Appliances	**VA Rating**
1. Water Heater	<u>3380</u> (Total 208V Appliance)

3380

APPLIANCES TOTAL   =   9860VA

APPLY DEMAND FACTOR (Applicable, when number of above appliances exceeds four (4) or more.)

(Appliances Total) <u>9860VA</u> x .75 =   7395VA

APPLIANCE LOADS

4.	LINE LOAD	NEUTRAL LOAD (Refer to condition)
	7395VA	4860VA

Condition
Because the water heater is a 208 volts line-to-line load it will not contribute to the neutral load as the other appliances and therefore must not be included. As a result, the NEUTRAL LOAD is equal to

9860VA - 3380VA x .75 = <u>4860VA</u>

5. Clothes Dryer (208/120V-1φ) <NEC and Table 220.54> (Use 5000W [VA] or nameplate rating, whichever is larger.)

CLOTHES DRYER

5.	LINE LOAD	NEUTRAL LOAD [NEC 220.61(C)(1)]
	5000VA	5000VA

6. Electric Ranges and Other Cooking Appliances (208/120V-1ϕ) <NEC and Table 220.55>

Minimum Demand Load determined per Columns B or C to Table 220.55.

Applying Column B

Demand Percent (2 appliances) = .65
(5.8kW) 5800W + (6.7kW) 6700W x .65 = 8125W

Applying Column C

Maximum Demand (2 appliances)
11kW(11,000W)

Use the smallest demand load of Column B (8125W) and Column C (11,000W).

REMINDER: Although it may appear that the initial provision of **Note 4** to Table 220.55 could have been applied toward the use of the two cooking appliances, it cannot. Because both cooking appliances are used toward calculating the service load for this occupancy such provision is not applicable. The provisions of **Note 4** to Table 220.55 are only applicable for branch-circuit loads and not for service and feeder load calculations. If the provision of **Note 4** to Table 220.55 could be applied in treating the two cooking appliances as one range, the provisions of **Note 1** to Table 220.55 would be required to obtain a demand load; but yet would not produce a demand load smaller than 8125W.

Based on the exclusive use of the WMO per Column B the neutral load is,

(5.8kW) 5800W x .80 = 4640W

RANGE/COOKING APPLIANCES

6.	LINE LOAD	NEUTRAL LOAD [NEC 220.61(C)(1)]
	8125VA	4640VA

7. Heating and Air-Conditioning Equipment <NEC 220.50, 220.51, 220.60, Table 430.248 and 440.6(A)> (Include VA rating of air handler [blower].)

Heating Load (larger)

18kW (18,000VA)

AC Load

    6384VA

          Electric Heat   Blower
Line Load = 18,000VA + 1416VA = 19,416VA

HEATING and AC

7.	LINE LOAD	NEUTRAL LOAD
	19,416VA	0

8. Largest Motor <NEC 430.17, 430.24, 440.7 and 440.33> (Use motor with highest full-load current [FLC] regardless of voltage rating.)

Largest Motor (LM) = 900VA (7.5A) disposal rated for 120 volts

900VA x .25 = 225VA
(LM)

LARGEST MOTOR

8.	LINE LOAD	NEUTRAL LOAD (120V motors only)
	225VA	225VA

TOTAL DEMAND LOAD (LINE and NEUTRAL) (Add lines 1. − 8.)

	LINE LOAD	NEUTRAL LOAD
1. - 3.	4995VA	4995VA
4.	7395VA	4860VA
5.	5000VA	5000VA
6.	8125VA	4640VA
7.	19,416VA	0
8.	225VA	225VA
	45,156VA	19,720VA

DWELLING'S OPERATING LINE VOLTAGE - 208V
(Given operating voltage or as determined per test examination)

CALCULATE MINIMUM LINE and NEUTRAL LOADS
(Divide Total Demand Load [VA] by operating line voltage [V])

LINE LOAD = 45,156VA / 208V = 217.1A
NEUTRAL LOAD = 19,720VA / 208V = 94.81A

SIZE SERVICE (Size of service based on the calculated **LINE LOAD**)

SIZE SERVICE REQUIRED (minimum) <u>225A</u>
(Single rating - Main overcurrent device **225A**)

SIZING FEEDER/SERVICE CONDUCTORS

Applying NEC 240.4(B) and Table 310.15(B)(16), as a minimum use 3/0 AWG copper conductors (75°C) which has an ampacity of 200A as the service conductors.

FEEDER/SERVICE CONDUCTORS <u>3/0 AWG copper conductors</u>

SIZING NEUTRAL CONDUCTOR(S)

Using Table 310.15(B)(16), as a minimum, use a 3 AWG copper conductor (75°C) which has an ampacity of 100A, as the neutral conductor. The use of a 3 AWG copper conductor as the neutral is in compliance with NEC 250.24(C)(1) and Table 250.66.

NEUTRAL CONDUCTOR(S) <u>3 AWG copper conductor</u>

88. Re-calculate question No. 87., if all loads are supplied by a 240/120V, single-phase system. Upon concluding compare the results of both single and three phase system's calculations.

Answers to question No. 88 consist of 36 pages (pages 136 – 172)

To satisfy question No. 88., the ratings of all existing three-phase loads were adjusted as follows:

Laundromat - 400SF

12 - Single-phase commercial washers - (21.7A - 240/120V)
6 - Single-phase commercial clothes dryers - (240/120V)
Heating Elements - 9.2kW, Motor - 25.6A

Swimming Pool

5HP pump motor (240V - 1φ)
2HP circulatory motor (240V - 1φ)

With these adjustments the house and apartment loads can be re-calculated using the same previously used methods as reference formats. When supplied by a 240/120V service, prohibited reductions of the neutrals loads are not applicable.

208V-3φ adjustments to 240V-1φ are highlighted.

HOUSE LOAD SINGLE-PHASE LOAD CALCULATIONS
AT 240/120V

For "House Load" reference format see **(Worksheet G - Volume 4)** STANDARD LOAD CALCULATION FOR NONDWELLING BUILDING (COMMERICIAL and` INDUSTRIAL).

HOUSE LOAD - 240/120V 1-phase (House Load calculation covers pages 137 – 142)

1. GENERAL LIGHTING or ACTUAL LIGHTING LOADS

General Lighting Load <NEC References - 220.12 and Table 220.12>

Office - 740SF x 3.5VA = 2590VA

   A. General Lighting Load = ~~2590VA~~ <Omit General Lighting Load - Actual Lighting Load Larger>

Actual Lighting Load

Type Fixture	VA rating	No. of Fixtures		TOTAL VA
Office				
Fluorescent	120V x 1.65A x	12	=	2376
Recessed Cans 100W(VA) x			=	500
B. Actual Lighting Load			=	2876

2876VA x 1.25 = 3595VA

1. LINE LOAD	NEUTRAL LOAD	
	Permitted Reduction	Prohibited Reduction
3595VA	2876VA	--

2. OTHER LIGHTING LOADS

   A. Sign/Outline (S/O) Lighting <NEC References - 220.12(F) and 600.5(A)>

      Signs (120V) - 2000VA x 3 = 6000VA

            Total (Sign/Outline Lighting) = 6000VA

   B. Outside Lighting <NEC Reference - 220.18(B)>

Metal Halide - 240V x 2.48A x 13 = 7737.6VA

Total (Outside Lighting) = 7737.6VA

C. NA

D. Track Lighting <NEC Reference - 220.43(B)> (Voltage rating -120V)

(8' ÷ 2') x 150VA = 600VA

E. Miscellaneous (Write-ins. List individual voltage rating of each lighting load.)

Underwater light fixtures - 120V x 1.23A x 8 = 1180.8VA
Laundromat, Exercise Room and Spa (Fluorescent) - 120V x 1.65A x 16 = 3168VA

Total (Miscellaneous)   = 4348.8VA

F. Other Lighting (LINE and NEUTRAL) Loads Total [Add lines (A.) - (E.)]

Other Lighting Loads Total = 18,686.40VA

TOTAL = 18,686.40VA x 1.25 = 23,358VA

2.   <u>LINE LOAD</u>   <u>NEUTRAL LOAD</u>
                                                              (120V)
                                                    Permitted   Prohibited
                                                    Reduction   Reduction

23,358VA   10,948.8VA   --

3. RECEPTACLE LOADS

A. Non-continuous duty (Other Than Bank or Office Building) <NEC 220.14(I)>

<u>Laundromat (5), Exercise Room and Spa (15)</u>

20 x 180VA = 3600VA

B. Non-continuous duty (Office Building) <NEC 220.14(K)>

<u>Office</u>

(1) 10 x 180VA = 1800VA (larger)

(2) 740SF x 1VA = 740VA

C. NA

D. Total Receptacle (LINE and NEUTRAL) Load

Total Receptacle loads - 3600VA + 1800VA = 5400VA

3.	LINE LOAD		NEUTRAL LOAD	
			Permitted Reduction	Prohibited Reduction
	5400VA		5400VA	--

4. KITCHEN EQUIPMENT - NA

5. SPECIFIC LOADS (Appliances, Computer Equipment, Office equipment, etc.)

Type Load	Calculation		
240V			
Copier	240V x 10.3A	=	2472VA
Dryer (Heating Elements) (6)	9200W x 6	=	55,200VA
Jacuzzi Heater (2)	3300W x 2	=	6600VA
Water Heater	4500W	=	4500VA
			68,772VA
120V			
Desktop computers (4)	120V x 1.3A x 4	=	624VA
Laser printers (2)	120V x 3.8A x 2	=	912VA
			1536VA
	TOTAL	=	70,308VA

5.	LINE LOAD		NEUTRAL LOAD	
			Permitted Reduction	Prohibited Reduction
	70,308VA		1536VA	--

6. MOTOR LOADS <NEC Reference - 220.50 >

A. Continuous Duty

Motor Load	Calculation		
120V			
¾HP Circulatory Pump (2)	120V x 13.8A* x 2	=	3312VA
240V			
5HP Swimming Pool Pump	240V x 28A*	=	6720VA
2HP Swimming Pool Pump	240V x 12A*	=	2880VA
*Table 430.248			12,912VA

B. Other Than Continuous Duty

Motor Load	Calculation		
120V			
Soft drink machine	120V x 10.4A	=	1248VA
240V			
Dryers (6)	240V x 25.6A x 6	=	36,864VA
Washers (12)	240V x 21.7A x 12	=	62,496VA
99,360VA			

C. and D. - NA

E. Total Motor Loads – [LINE LOAD - Add lines (A.) and (B.)] – [NEUTRAL LOAD - Total motor loads with neutral connections (120V)]

$$\text{TOTAL} = 113,520\text{VA}$$

6. | LINE LOAD | NEUTRAL LOAD |

	Permitted Reduction	Prohibited Reduction
113,520VA	4560VA	--

7. MEDICAL EQUIPMENT - NA

8. INDUSTRIAL EQUIPMENT - NA

9. HEATING and **AIR-CONDITIONING (AC) EQUIPMENT** <NEC References 220.50, 220.51, 220.60, 430.6(A)(1) and 440.6(A)>

ELECTRICAL HEATING UNITS

Heating Unit	Calculation
Furnaces - 13.3kW (2)	13,300W x 2 = 26,600W (VA)
Furnace - 8kW	= 8000W (VA)
	TOTAL HEAT = 34,600VA

AIR-CONDITIONING (AC) UNITS

AC Unit	Calculation
Units - 2743VA (2)	2743VA x 2 = 5486VA
Unit - 1645VA	= 1645VA
	TOTAL AC = 7131VA

Line Load $=$ 34,600VA $+$ (4488VA) $=$ 39,088VA
       (Largest Load)   (Blowers)

¾ HP — 240V x 6.9A* x 2 =   3312VA
½ HP — 240V x 4.9A*   =   1176VA
*Table 430.248   4488VA

9.   LINE LOAD   NEUTRAL LOAD
                                          Permitted   Prohibited
                                          Reduction   Reduction

       39,088VA   0   --

10. LARGEST MOTOR

5HP Swimming Pool Motor (motor with highest current per NEC 430.17 and 440.7)
6720VA x .25(25 percent) = 1680VA

10.   LINE LOAD   NEUTRAL LOAD
                                          Permitted   Prohibited
                                          Reduction   Reduction

       1680VA   0   0

TOTAL DEMAND LOAD (LINE and NEUTRAL) (List each computed line and neutral loads below and total lines 1. – 10.) Highlights identify loads affected by re-calculations.

	LINE LOAD	NEUTRAL LOAD Permitted Reduction	Prohibited Reduction
1. General Lighting	3595VA	2876.0VA	--
2. Other Lighting Loads	23,358VA	10,948.8VA	--
3. Receptacle Loads	5400VA	5400.0VA	--
4. Kitchen Equipment	--	--	--
5. Specific Loads	70,308VA	1536.0VA	--
6. Motor Loads	113,520VA	4560.0VA	--
7. Medical Equipment	--	--	--
8. Industrial Equipment	--	--	--
9. Heating and AC Equip.	39,088VA	0	--
10. Largest Motor	**1680VA**	**0**	--
TOTAL =	256,949VA	25,320.8VA	--

BUILDING'S OPERATING LINE VOLTAGE - 240V
(Given operating voltage or as determined per test examination)

CALCULATE MINIMUM LINE AND NEUTRAL LOADS
(Divide Total Demand Load **[VA]** by operating line voltage **[V]**)

LINE LOAD = 256,949VA / 240V = 1070.62A

NEUTRAL LOAD
   Permitted = 25,320.8VA / 240V = 105.5A

   Total Neutral Load = 105.5A

SIZE SERVICE (Size of service based on the calculated LINE LOAD)

SIZE SERVICE REQUIRED (minimum) 1200A
(Single rating or combination - Main overcurrent device(s) to total 1200A)

The house service load is 1070.62A (1φ) and the neutral load is 105.5A.

SINGLE-PHASE LOAD CALCULATIONS PER APARTMENT COMPLEX AT 240/120V

For "Multifamily Dwelling" reference format see **Worksheet C** (Volume 4) – STANDARD LOAD CALCULATIONS FOR MULTIFAMILY DWELLING.

To re-calculate the loads of each arrangement of apartment units only those equipment loads with dual voltage/wattage (VA) ratings are required for re-calculation. All 120V load calculations can be applied as is.

54 units - 240/120V 1-phase (54 units load calculation covers pages 142 – 145)

4. Appliance Loads (Fastened-In-Place) <NEC 220.53>

120V Appliances	VA	No. of Appliances	Total VA
1. Dishwasher	900	32	28,800
2. Dishwashers	1250	22	27,500
3. Disposals	600	32	19,200
4. Disposals	900	22	19,800
5. Trash compactors	780	22	17,160
6. Microwaves	1000	15	15,000
7. Microwaves	1200	17	20,400
8. Microwaves	1600	12	19,200
9. Microwaves	1800	10	18,000
10. Heat-Vent-Lights	1300	15	19,500
11. Heat-Vent-Lights	1600	29	46,400
12. Heat-Vent-Lights	1750	10	17,500
		(Total 120V Appliances)	268,460

240V Appliances	VA	No. Appliances	Total VA
1. Water Heaters	4500	54	243,000
		(Total 240V Appliances)	243,000

APPLIANCES TOTAL   = 511,460VA

APPLY DEMAND FACTOR (Applicable, when number of above appliances exceeds four (4) or more.)

(Appliances Total) 511,460VA x .75 = 383,595VA

APPLIANCE LOADS

4. LINE LOAD   NEUTRAL LOAD (Refer to condition)

    383,595VA   201,345VA

Condition

Because the water heaters are 240 volts line-to-line loads they will not contribute to the neutral load as the other appliances and therefore must not be included. As a result, the NEUTRAL LOAD is equal to

511,460VA − 243,000 x .75 = 201,345VA

5. Clothes Dryers[2] <NEC and Table 220.54>

[2]650SF units (15) not equipped with washer/dryer connections.

Number of dryers = 39

Per Table 220.54

the demand factor (%) = 35% − [.5% x (39 − 23)] = 27% (.27)

The demand load of dryers

5000VA x 39 x .27 = 52,650VA

CLOTHES DRYERS

5. LINE LOAD   NEUTRAL LOAD [NEC 220.61(B)(1)]

    52,650VA   36,855VA (52,650VA x .70 of line load)

6. Electric Ranges and Other Cooking Appliances <NEC and Table 220.55>

LINE LOAD [240/120V-Range (R)/Wall Mounted Oven (WMO) and 240V-Cooktop (CT)]

Because the ratings of the cooking appliances all fall under the provisions of Column C, Column C will be used exclusively to determine the demand load of the appliances. Although most of the appliances could be calculated applying the provisions of Column B and the remaining appliances applying Column C and then total, this approach would not only prove to be impractical but would also yield a larger demand load.

Number of appliances = 76 (Appliances rated between 7.18kW – 10.4kW)

Appliances maximum demand load per Column C

76 appliances = 25kW + (.75kW x 76) = 82kW (82,000VA)

NEUTRAL LOAD – 240/120V (Range(R)/Wall Mounted Oven (WMO)

Appliances maximum demand load per Column C

54 appliances = 25kW + (.75kW x 54) = 65.5kW (65,500VA)

ELECTRIC RANGES

6.	LINE LOAD	NEUTRAL LOAD [NEC 220.61(B)(1)]
	82,000VA	45,850VA (65,500VA x .70 of line load)

7. Heating and Air-Conditioning (AC) Equipment <NEC References - 220.50, 220.51, 220.60, Table 430. 248 and 440.6(A)> (Include VA rating of air handler [blower].)

Heating Load (larger)
    Electric Heat
        13.3kW (13,300VA) x <u>15</u> = 199,500VA
        16kW (16,000VA) x <u>17</u> = 272,000VA
        20kW (20,000VA) x <u>12</u> = 240,000VA
        24kW (24,000VA) x <u>10</u> = <u>240,000VA</u>
                                951,500VA

AC Load

    2743VA x <u>15</u> = 41,145VA
    3429VA x <u>17</u> = 58,293VA
    4114VA x <u>12</u> = 49,368VA
    4800VA x <u>10</u> = <u>48,000VA</u>
                  196,806VA

$$\text{Line Load} = \underset{\text{Electric Heat}}{\underline{951,500VA}} + \underset{\text{Blowers (1)}}{\underline{785VA}} \times \underline{15} + \underset{\text{Blowers (2)}}{\underline{884VA}} \times \underline{17} + \underset{\text{Blowers (3)}}{\underline{992VA}} \times \underline{12}$$
$$+ \underset{\text{Blowers (4)}}{\underline{1065VA}} \times \underline{10} = 1,000,857VA$$

HEATING and AC

7.	LINE LOAD	NEUTRAL LOAD
	1,000,857VA	0

TOTAL DEMAND LOAD (LINE AND NEUTRAL) (Add lines 1. − 8.) Highlights identify loads affected by re-calculations.

	LINE LOAD	NEUTRAL LOAD
1. - 3.	109,162.5VA	109,162.5VA
4.	383,595.0VA	201,345.0VA
5.	52,650.0VA	36,855.0VA
6.	82,000.0VA	45,850.0VA
7.	1,000,857.0VA	0
8.	225.0VA	225.0VA
	1,628,489.5VA	393,437.5VA

DWELLING'S OPERATING LINE VOLTAGE - 240V
(Given operating voltage or as determined per test examination)

CALCULATE MINIMUM LINE AND NEUTRAL LOADS
(Divide Total Demand Load [VA] by operating line voltage [V])

LINE LOAD = $\underline{1,628,489.5}VA / \underline{240}V = \underline{6785.37}A$
NEUTRAL LOAD = $\underline{393,437.5}VA / \underline{240}V = \underline{\text{1639.32A}} \: 1207.52A$

> *Where the neutral load exceeds 200A, NEC 220.61(B)(2) permits the load to be reduced by 70 percent. Complete the following to determine permitted Neutral Demand Load.

(1) $\underset{\text{(Neutral Load)*}}{\underline{1639.32}}$ A − 200A = $\underset{\text{(Remainder)}}{\underline{1439.32}A}$ x .70 = $\underset{\text{(Permitted Reduction)}}{\underline{1007.52}}$ A

(2) $\underset{\text{(Permitted Reduction)}}{\underline{1007.52}}$ A + 200A = $\underset{\text{(Permitted Neutral Demand Load)}}{\underline{1207.52}}$ A

SIZE SERVICE (Size of service based on the calculated **LINE LOAD**)

SIZE SERVICE REQUIRED (maximum) <u>7000A</u>
(Combination - Main overcurrent device(s) to total **7000A**)

The overall apartment complex line load is 6785.37A (1φ) and the neutral load is 1207.52A.

SINGLE-PHASE LOAD CALCULATIONS
PER APARTMENT UNITS AT 240/120V

For "Multifamily Dwelling" reference format see **Worksheet C** (Volume 4) – STANDARD LOAD CALCULATIONS FOR MULTIFAMILY DWELLING.

650SF units (15) - 240/120V 1-phase (650SF units load calculation covers pages 146 – 149)

4. Appliance Loads (Fastened-In-Place) <NEC 220.53>

120V Appliances	VA	No. of Appliances	Total VA
1. Dishwashers	900	15	13,500
2. Disposals	600	15	9,000
3. Microwaves	1000	15	15,000
4. Heat-Vent-Lights	1300	15	19,500
		(Total 120V Appliances)	57,000

240V Appliances	VA	No. Appliances	Total VA
1. Water Heaters	4500	15	67,500
		(Total 240V Appliances)	67,500

APPLIANCES TOTAL   = 124,500VA

APPLY DEMAND FACTOR (Applicable, when number of above appliances exceeds four (4) or more.)

(Appliances Total) 124,500VA x .75 = 93,375VA

APPLIANCE LOADS

4.   LINE LOAD   NEUTRAL LOAD (Refer to condition)

93,375VA   42,750VA

Condition
Because the water heaters are 240 volts line-to-line loads they will not contribute to the neutral load as the other appliances and therefore must not be included. As a result, the NEUTRAL LOAD is equal to

124,500VA – 67,500VA x .75 = 42,750VA

6. Electric Ranges and Other Cooking Appliances <NEC and Table 220.55>

Number of ranges = 15   Ranges demand load calculated per Column B

8.1kW x 15 = 121.5kW [121,500VA])

Demand Factor per Column B

15 ranges = .32

$$121,500VA \times .32 = 38,880VA$$

Appliances maximum demand load per Column C

$$15 \text{ appliances} = 30kW \ (30,000VA)$$

Compared to the Maximum Demand of Column C (30,000VA) - Use 30,000VA as the demand load for appliances.

ELECTRIC RANGES

6.	LINE LOAD	NEUTRAL LOAD [NEC 220.61(B)(1)]
	30,000VA	21,000VA (30,000VA x .70 of line load)

7. Heating and Air-Conditioning (AC) Equipment <NEC References - 220.50, 220.51, 220.60, Table 430. 248 and 440.6(A)> (Include VA rating of air handler [blower].)

Heating Load

Electric Heat
13.3kW (13,300VA) x 15 = 199,500VA

AC Load

2743VA x 15 = 41,145VA

Heating load is larger.

Electric Heat   Blowers
Line Load = 199,500VA + 785VA x 15 = 211,275VA

HEATING and AC

7.	LINE LOAD	NEUTRAL LOAD
	211,275VA	0

TOTAL DEMAND LOAD (LINE and **NEUTRAL)** (Add lines 1. − 8.) Highlights identify loads affected by re-calculations.

	LINE LOAD	NEUTRAL LOAD
1. - 3.	27,937.5VA	27,937.50VA
4.	93,375.0VA	42,750.00VA
5.	0	0
6.	30,000.0VA	21,000.00VA
7.	211,275.0VA	0
8.	150.0VA	150.00VA (Largest Motor – 25% of 600VA disposal)
	362,737.5VA	91,837.50VA

DWELLING'S OPERATING LINE VOLTAGE - 240V
(Given operating voltage or as determined per test examination)

CALCULATE MINIMUM LINE and **NEUTRAL LOADS**
(Divide Total Demand Load **[VA]** by operating line voltage **[V]**)

LINE LOAD = 362,737.5VA / 240V = 1511.41A
NEUTRAL LOAD = 91,837.5VA / 240V = ~~382.66A~~* 327.86A

*Where the neutral load exceeds 200A, NEC 220.61(B)(2) permits the load to be reduced by 70 percent. Complete the following to determine permitted Neutral Demand Load.

(1) 382.66   A – 200A = 182.66 A x .70 = 127.86   A
   (Neutral Load)*   (Remainder)   (Permitted Reduction)

(2) 127.86   A + 200A = 327.86   A
   (Permitted Reduction)   (Permitted Neutral Load)

SIZE SERVICE (Size of service based on the calculated **LINE LOAD**)

SIZE SERVICE REQUIRED (minimum) **1600A**
(Single rating or combination - Main overcurrent device(s) to total **1600A**)

The conductors or combinations selected for this installation are only limited for this question. Conductor ampacity derating is not applied in sizing the service and neutral conductors.

SIZING FEEDER/SERVICE CONDUCTORS

Using NEC 310.10(H)(1) and Table 310.15(B)(16), as a minimum based on NEC 240.4(C), use four 600 kcmil copper conductors (75°C) per phase which have an individual ampacity of 420A, as the service conductors.

FEEDER/SERVICE CONDUCTORS 4-600 kcmil copper conductors per line

SIZING NEUTRAL CONDUCTOR(S)

Based on the calculated neutral load and the use of parallel service conductors, the neutral conductors must also be installed in parallel per NEC 310.10(H)(1). Because the service conductors will consist of four parallel conductors per phase, the neutral conductors must also consist of four parallel conductors based on the calculated neutral load being divided by four which results to 81.97A (327.86A / 4). Per Table 310.15(B)(16), at 75°C, a 4 AWG copper conductor which has a rated ampacity of 85A will satisfy the calculated neutral load being distributed amongst four such conductors. However, because NEC 310.10(H)(1) only allows a 1/0 AWG conductor or larger to be installed in parallel, four 1/0 AWG copper conductors must be used instead as the parallel neutral conductors.

In total circular mils, the four 1/0 AWG parallel neutral conductors will exceed the required 3/0 AWG grounding electrode conductor, when considered. The size of the grounding electrode conductor is based on the equivalent area of the parallel service conductors [2400 kcmil (600 kcmil x 4)] per Table 250.66 which is over 1100 kcmil copper.

NEUTRAL CONDUCTOR(S) 4-1/0 AWG copper conductors

See question Nos. 1. - 8. of Article 250 for further reference when the grounded (neutral) conductor is not used as a circuit conductor.

900SF units (17) - 240/120V 1-phase (650SF units load calculation covers pages 149 – 152)

4. Appliance Loads (Fastened-In-Place) <NEC 220.53>

120V Appliances	VA	No. of Appliances	Total VA
1. Dishwasher	900	17	15,300
2. Disposal	600	17	10,200
3. Microwave	1200	17	20,400
4. Heat-Vent-Light	1600	17	27,200
		(Total 120V Appliances)	73,100

240V Appliances	VA	No. Appliances	Total VA
1. Water Heater	4500	17	76,500
		(Total 240V Appliances)	76,500

APPLIANCES TOTAL = 149,600VA

APPLY DEMAND FACTOR (Applicable, when number of above appliances exceeds four (4) or more.)

(Appliances Total) 149,600VA x .75 = 112,200VA

APPLIANCE LOADS

4.   <u>LINE LOAD</u>   <u>NEUTRAL LOAD (Refer to condition)</u>

      112,200VA   54,825VA

<u>Condition</u>
Because the water heaters are 240 volts line-to-line loads they contribute to the neutral load as the other appliances and therefore must not be included. As a result, the NEUTRAL LOAD is equal to

$$149,600VA - 76,500VA \times .75 = 54,825VA$$

5. Clothes Dryers <NEC and Table 220.54>

Number of dryers = 17

Per Table 220.54

$$\text{the demand factor (\%)} = 47\% - [1\% \times (17 - 11)] = 41\%(.41)$$

The demand load of dryers

$$5000VA \times 17 \times .41 = 34,850VA$$

CLOTHES DRYERS

5.   <u>LINE LOAD</u>   <u>NEUTRAL LOAD [NEC 220.61(B)(1)]</u>

      34,850VA   24,395VA (34,850VA × .70 of line load)

6. Electric Ranges and Other Cooking Appliances <NEC and Table 220.55>

Appliances maximum demand load per Column C

$$17 \text{ appliances} = 32kW (32,000VA)$$

ELECTRIC RANGES

6.   <u>LINE LOAD</u>   <u>NEUTRAL LOAD [NEC 220.61(B)(1)]</u>

      32,000VA   22,400VA (32,000VA × .70 of line load)

7. Heating and Air-Conditioning (AC) Equipment <NEC References - 220.50, 220.51, 220.60, Table 430. 248 and 440.6(A)> (Include VA rating of air handler [blower].)

Heating Load (larger)

Electric Heat
16kW (16,000VA) x <u>17</u> = 272,000VA

AC Load - 3429VA x <u>17</u> = 58,293VA

Electric Heat   Blowers
Line Load = <u>272,000VA</u> + <u>884VA</u> x <u>17</u> = <u>287,028VA</u>

HEATING and AC

7.   <u>LINE LOAD</u>   <u>NEUTRAL LOAD</u>

287,028VA   0

TOTAL DEMAND LOAD (LINE and **NEUTRAL)** (Add lines 1. – 8.) Highlights identify loads affected by re-calculations.

<u>LINE LOAD</u>	<u>NEUTRAL LOAD</u>
1. - 3. 44,550VA	44,550VA
4. 112,200VA	54,825VA
5. 34,850VA	24,395VA
6. 32,000VA	22,400VA
7. 287,028VA	0
8. 150VA	150VA (Largest Motor – 25% of 600VA disposal)
510,778VA	146,320VA

DWELLING'S OPERATING LINE VOLTAGE - <u>240V</u>
(Given operating voltage or as determined per test examination)

CALCULATE MINIMUM LINE and **NEUTRAL LOADS**
(Divide Total Demand Load **[VA]** by operating line voltage **[V]**)

LINE LOAD = <u>510,778</u>VA / <u>240</u>V = <u>2128.24</u>A
NEUTRAL LOAD = <u>146,320</u>VA / <u>240</u>V = ~~609.67~~A* <u>486.77</u>A

*Where the neutral load exceeds 200A, NEC 220.61(B)(2) permits the load to be reduced by 70 percent. Complete the following to determine permitted Neutral Demand Load.

(1) <u> 609.67 </u>A – 200A = <u> 409.04 </u>A x .70 = <u> 286.77 </u>A
        (Neutral Load)*   (Remainder)   (Permitted Reduction)

(2) <u> 286.77 </u>A + 200A = <u> 486.77 </u>A
        (Permitted Reduction)   (Permitted Neutral Load)

SIZE SERVICE (Size of service based on the calculated **LINE LOAD**)

SIZE SERVICE REQUIRED (minimum) **2500A**
(Single rating or combination - Main overcurrent device(s) to total **2500A**)

The conductors or combinations selected for this installation are only limited for this question. Conductor ampacity derating is not applied in sizing the service and neutral conductors.

SIZING FEEDER/SERVICE CONDUCTORS

Using NEC 310.10(H)(1) and Table 310.15(B)(16), as a minimum based on NEC 240.4(C), use four 1500 kcmil copper conductors (75°C) per phase which have an individual ampacity of 625A, as the service conductors.

FEEDER/SERVICE CONDUCTORS 4-1500 kcmil copper conductors per line

SIZING NEUTRAL CONDUCTOR(S)

Based on the calculated neutral load and the use of parallel service conductors, the neutral conductors must also be installed in parallel per NEC 310.10(H)(1). Because the service conductors will consist of four parallel conductors per phase, the neutral conductors must also consist of four parallel conductors based on the calculated neutral load being divided by four which results to 121.69A (486.77A / 4). Per Table 310.15(B)(16), at 75°C, a 1 AWG copper conductor which has a rated ampacity of 130A will satisfy the calculated neutral load being distributed amongst four such conductors. However, because NEC 310.10(H)(1) only allows a 1/0 AWG conductor or larger to be installed in parallel, four 1/0 AWG copper conductors must be used instead as the parallel neutral conductors.

In total circular mils, the four 1/0 AWG parallel neutral conductors will exceed the required 3/0 AWG grounding electrode conductor, when considered. The size of the grounding electrode conductor is based on the equivalent area of the parallel service conductors [6000 kcmil (1500 kcmil x 4)] per Table 250.66 which is over 1100 kcmil copper.

NEUTRAL CONDUCTOR(S) 4-1/0 AWG copper conductors

1200SF units (12) - 240/120V 1-phase (1200SF units load calculation covers pages 152 – 156)

4. Appliance Loads (Fastened-In-Place) <NEC 220.53>

120V Appliances	VA	No. of Appliances	Total VA
1. Dishwasher	1250	12	15,000
2. Disposal	900	12	10,800
3. Trash Compactor	780	12	9,360
4. Microwave	1600	12	19,200
5. Heat-Vent-Light	1600	12	19,200
	(Total 120V Appliances)		73,560

240V Appliances	VA	No. Appliances	Total VA
1. Water Heater	4500	12	54,000
		(Total 240V Appliances)	54,000

APPLIANCES TOTAL   = 127,560VA

APPLY DEMAND FACTOR (Applicable, when number of above appliances exceeds four (4) or more.)

(Appliances Total) 127,560VA x .75 = 95,670VA

APPLIANCE LOADS

4. LINE LOAD	NEUTRAL LOAD (Refer to condition)
95,670VA	55,170VA

Condition
Because the water heaters are 240 volts line-to-line loads they will not contribute to the neutral load as the other appliances and therefore must not be included. As a result, the NEUTRAL LOAD is equal to

127,560VA – 54,000VA x .75 = 55,170VA

5. Clothes Dryers <NEC and Table 220.54>

Number of dryers = 12

Per Table 220.54

the demand factor (%) = 47% – [1% x (12 – 11)] = 46%(.46)

The demand load of dryers

5000VA x 12 x .46 = 27,600VA

CLOTHES DRYERS

5. LINE LOAD	NEUTRAL LOAD [NEC 220.61(B)(1)]
27,600VA	19,320VA (27,600VA x .70 of line load)

6. Electric Ranges and Other Cooking Appliances <NEC and Table 220.55>

LINE LOAD - [240/120V Wall Mounted Oven (WMO) and 240V Cooktop (CT)]

Number of appliances = 24 (Appliances rated between 7.18kW – 8.4kW)

Appliances demand load calculated per Column B

$$7.18kW \times 12 = \quad 86.16kW \text{ (WMO)}$$
$$8.4kW \quad \times 12 = \quad \underline{100.80kW} \text{ (CT)}$$
$$186.96kW \text{ (186,960W [VA])}$$

Demand Factor per Column B

24 appliances = .26

$$186,960VA \times .26 = 48,609.6VA$$

Appliances maximum demand load per Column C

$$24 \text{ appliances} = 39kW \text{ (39,000VA)}$$

Compared to the Maximum Demand of Column C (39,000VA) - Use 39,000VA as the demand load for appliances.

NEUTRAL LOAD - 240/120V Wall Mounted Oven (WMO)

Number of appliances (with neutral connections) = 12

Appliance demand load calculated per Column B

$$7.18kW \times 12 = 86.16kW \text{ (861,600VA) (WMO)}$$

Demand Factor per Column B

12 WMO = .32

$$861,600VA \times .32 = 27,571.2VA$$

Appliances maximum demand load per Column C

$$12 \text{ WMO} = 27kW \text{ (27,000VA)}$$

Compared to the Maximum Demand of Column C (27,000VA) - Use 27,000VA as the demand load for appliances.

ELECTRIC RANGES

6.	LINE LOAD	NEUTRAL LOAD [NEC 220.61(B)(1)]
	39,000VA	18,900VA (27,000VA x .70 of line load)

7. Heating and Air-Conditioning (AC) Equipment <NEC References - 220.50, 220.51, 220.60, Table 430. 248 and 440.6(A)> (Include VA rating of air handler [blower].)

Heating Load

Electric Heat
20kW (20,000VA) x <u>12</u> = 240,000VA

AC Load

4114VA x <u>12</u> = 49,368VA

Heating load is larger

Line Load = $\underset{\text{Electric Heat}}{\underline{240,000\text{VA}}}$ + $\underset{\text{Blowers}}{\underline{992\text{VA}}}$ x <u>12</u> = <u>251,904VA</u>

HEATING and AC

7.	LINE LOAD	NEUTRAL LOAD
	251,904VA	0

TOTAL DEMAND LOAD (LINE and NEUTRAL) (Add lines 1. – 8.) Highlights identify loads affected by re-calculations.

	LINE LOAD	NEUTRAL LOAD
1. - 3.	35,970VA	35,970VA
4.	95,670VA	55,170VA
5.	27,600VA	19,320VA
6.	39,000VA	18,900VA
7.	251,904VA	0
8.	225VA	225VA
	450,369VA	129,585VA

DWELLING'S OPERATING LINE VOLTAGE - <u>240</u>V
(Given operating voltage or as determined per test examination)

CALCULATE MINIMUM LINE and NEUTRAL LOADS
(Divide Total Demand Load [VA] by operating line voltage [V])

LINE LOAD = 450,369VA / 240V = 1876.54A
NEUTRAL LOAD = 129,585VA / 240V = ~~539.94A~~* 437.96A

*Where the neutral load exceeds 200A, NEC 220.61(B)(1) permits the load to be reduced by 70 percent. Complete the following to determine permitted Neutral Demand Load.

(1) <u>539.94</u> A – 200A = <u>339.94</u> A x .70 = <u>237.96</u> A
    (Neutral Load)*   (Remainder)   (Permitted Reduction)

(2) <u>237.96</u> A + 200A = <u>437.96</u> A
    (Permitted Reduction)   (Permitted Neutral Demand Load)

SIZE SERVICE (Size of service based on the calculated **LINE LOAD**)

SIZE SERVICE REQUIRED (minimum) 2000A
(Single rating or combination - Main overcurrent device(s) to total **2000A**)

The conductors or combinations selected for this installation are only limited for this question. Conductor ampacity derating is not applied in sizing the service and neutral conductors.

SIZING FEEDER/SERVICE CONDUCTORS

Using NEC 310.10(H)(1) and Table 310.15(B)(16), as a minimum based on NEC 240.4(C), use four 900 kcmil copper conductors (75°C) per phase which have an individual ampacity of 520A, as the service conductors

FEEDER/SERVICE CONDUCTORS 4-900 kcmil copper conductors per phase

SIZING NEUTRAL CONDUCTOR(S)

Based on the calculated neutral load and the use of parallel service conductors, the neutral conductors must also be installed in parallel per NEC 310.10(H)(1). Because the service conductors will consist of four parallel conductors per phase, the neutral conductors must also consist of four parallel conductors based on the calculated neutral load being divided by four which results to 109.49A (437.96A / 4). Per Table 310.15(B)(16), at 75°C, a 2 AWG copper conductor which has a rated ampacity of 115A will satisfy the calculated neutral load being distributed amongst four such conductors. However, because NEC 310.10(H)(1) only allows a 1/0 AWG conductor or larger to be installed in parallel, four 1/0 AWG copper conductors must be used instead as the parallel neutral conductors.

In total circular mils, the four 1/0 AWG parallel neutral conductors will exceed the required 3/0 AWG grounding electrode conductor, when considered. The size of the grounding electrode conductor is based on the equivalent area of the parallel service conductors [3600 kcmil (900 kcmil x 4)] per Table 250.66 which is over 1100 kcmil copper.

NEUTRAL CONDUCTOR(S) 4-1/0 AWG copper conductors

1400SF units (10) - 240/120V 1-phase (1400SF units load calculation covers pages 157 – 160)

4. Appliance Loads (Fastened-In-Place) <NEC 220.53>

120V Appliances	VA	No. of Appliances	Total VA
1. Dishwasher	1250	10	12,500
2. Disposal	900	10	9000
3. Trash Compactor	780	10	7800
4. Microwave	1800	10	18,000
5. Heat-Vent-Light	1750	10	17,500
		(Total 120V Appliances)	64,800

240V Appliances	VA	No. Appliances	Total VA
1. Water Heater	4500	10	45,000
		(Total 240V Appliances)	45,000

APPLIANCES TOTAL　　　=　109,800VA

APPLY DEMAND FACTOR (Applicable, when number of above appliances exceeds four (4) or more)

(Appliances Total) 109,800VA x .75 = 82,350VA

APPLIANCE LOADS

4.　LINE LOAD	NEUTRAL LOAD (Refer to condition)
82,350VA	48,600VA

Condition

Because the water heaters are 240 volts line-to-line loads they will not contribute to the neutral load as the other appliances and therefore must not be included. As a result, the NEUTRAL LOAD is equal to

109,800VA – 45,000VA x .75 = 48,600VA

5. Clothes Dryers[2] <NEC and Table 220.54>

Number of dryers = 10

Per Table 220.54, the demand factor (%) = 50(.50)

The demand load of dryers

5000VA x 10 x .50 = 25,000VA

CLOTHES DRYERS

5. <u>LINE LOAD</u>   <u>NEUTRAL LOAD [NEC 220.61(B)(1)]</u>

   25,000VA   17,500VA (25,000VA x .70 of line load)

6. **Electric Ranges and Other Cooking Appliances** <NEC and Table 220.55>

Again, because the ratings of the cooking appliances all fall under the provisions of Column C, Column C will be used exclusively to determine the demand load of the appliances.

Number of appliances = 20 (Appliances rated between 7.7kW – 8.9kW)

Appliances maximum demand load per Column C

20 appliances = 35kW (35,000VA)

NEUTRAL LOAD – 240/120V (Wall Mounted Oven (WMO)

Appliances maximum demand load per Column C

10 appliances = 25kW (25,000VA)

ELECTRIC RANGES

6. <u>LINE LOAD</u>   <u>NEUTRAL LOAD [NEC 220.61(B)(1)]</u>

   35,000VA   17,500VA (25,000VA x .70 of line load)

7. **Heating and Air-Conditioning (AC) Equipment** <NEC References - 220.50, 220.51, 220.60, Table 430. 248 and 440.6(A)> (Include VA rating of air handler [blower].)

Heating Load

24kW (24,000VA) x <u>10</u> = 240,000VA

AC Load

4800VA x <u>10</u> = 48,000VA

Heating load is larger.

Electric Heat   Blowers
Line Load = <u>240,000VA</u> + <u>1065VA</u> x <u>10</u> = <u>250,650VA</u>

HEATING and AC

7.	LINE LOAD	NEUTRAL LOAD
	250,650VA	0

TOTAL DEMAND LOAD (LINE and NEUTRAL) (Add lines 1. – 8.) Highlights identify loads affected by re-calculations.

	LINE LOAD	NEUTRAL LOAD
1. - 3.	32,400VA	32,400VA
4.	82,350VA	48,600VA
5.	25,000VA	17,500VA
6.	35,000VA	17,500VA
7.	250,650VA	0
8.	225VA	225VA
	425,625VA	116,225VA

DWELLING'S OPERATING LINE VOLTAGE - 240V
(Given operating voltage or as determined per test examination)

CALCULATE MINIMUM LINE and NEUTRAL LOADS
(Divide Total Demand Load [VA] by operating line voltage [V])

LINE LOAD = 425,625VA / 240V = 1773.44A
NEUTRAL LOAD = 116,225VA / 240V = 484.27A* 398.99A

> *Where the neutral load exceeds 200A, NEC 220.61(B)(2) permits the load to be reduced by 70 percent. Complete the following to determine permitted Neutral Demand Load.

(1) $\underline{484.27}$ A – 200A = $\underline{284.27}$ A x .70 = $\underline{198.99}$ A
$\quad\quad$ (Neutral Load)* $\quad\quad\quad$ (Remainder) $\quad\quad\quad$ (Permitted Reduction)

(2) $\underline{198.99}$ A + 200A = $\underline{398.99}$ A
$\quad\quad$ (Permitted Reduction) $\quad\quad\quad$ (Permitted Neutral Demand Load)

SIZE SERVICE (Size of service based on the calculated **LINE LOAD**)

SIZE SERVICE REQUIRED (minimum) 2000**A**
(Single rating or combination - Main overcurrent device(s) to total **2000A**)

The conductors or combinations selected for this installation are only limited for this question. Conductor ampacity derating is not applied in sizing the service and neutral conductors.

SIZING FEEDER/SERVICE CONDUCTORS

Using NEC 310.10(H)(1) and Table 310.15(B)(16), as a minimum based on NEC 240.4(C), use four 900 kcmil copper conductors (75°C) per phase which have an individual ampacity of 520A, as the service conductors.

FEEDER/SERVICE CONDUCTORS 4-900 kcmil copper conductors per line

SIZING NEUTRAL CONDUCTOR(S)

Based on the calculated neutral load and the use of parallel service conductors, the neutral conductors must also be installed in parallel per NEC 310.10(H)(1). Because the service conductors will consist of four parallel conductors per phase, the neutral conductors must also consist of four parallel conductors based on the calculated neutral load being divided by four which results to 99.75A (398.99A / 4). Per Table 310.15(B)(16), at 75°C, a 3 AWG copper conductor which has a rated ampacity of 100A will satisfy the calculated neutral load being distributed amongst four such conductors. However, because NEC 310.10(H)(1) only allows a 1/0 AWG conductor or larger to be installed in parallel, four 1/0 AWG copper conductors must be used instead as the parallel neutral conductors.

In total circular mils, the four 1/0 AWG parallel neutral conductors will exceed the required 3/0 AWG grounding electrode conductor, when considered. The size of the grounding electrode conductor is based on the equivalent area of the parallel service conductors [3600 kcmil (900 kcmil x 4)] per Table 250.66 which is over 1100 kcmil copper.

NEUTRAL CONDUCTOR(S) 4-1/0 AWG copper conductors

SINGLE-PHASE LOAD CALCULATIONS PER INDIVIDUAL UNITS AT 240/120V

For "Multifamily Dwelling" reference format see **Worksheet A** (Volume 4) – STANDARD LOAD CALCULATIONS FOR ONE-FAMILY DWELLING.

650SF unit - 240/120V 1-phase (650SF unit load calculation covers pages 160 – 163)

4. Appliance Loads (Fastened-In-Place) <NEC 220.53>

120V Appliances	VA Rating
1. Dishwasher	900
2. Disposal	600
3. Microwave	1000
4. Heat-Vent-Light	1300
	3800 (Total 120V Appliances)

240V Appliances	**VA Rating**
1. Water Heater	<u>4500</u> (Total 208V Appliance)
	4500

APPLIANCES TOTAL   =   8300VA

APPLY DEMAND FACTOR (Applicable, when number of above appliances exceeds four (4) or more)

(Appliances Total) <u>8300VA</u> x .75 =   6225VA

APPLIANCE LOADS

4.	LINE LOAD	NEUTRAL LOAD (Refer to condition)
	6225VA	2850VA

<u>Condition</u>

Because the water heater is a 240 volts line-to-line load it will not contribute to the neutral load as the other appliances and therefore must not be included. As a result, the NEUTRAL LOAD is equal to

$$8300VA - 4500VA \times .75 = 2850VA$$

5. Clothes Dryer[2] (240/120V-1ϕ) <NEC and Table 220.54>

[2]650SF unit not equipped with washer/dryer connections.

CLOTHES DRYER

5.	LINE LOAD	NEUTRAL LOAD
	0	0

6. Electric Ranges and Other Cooking Appliances (240/120V-1ϕ) <NEC and Table220.55>

Minimum Demand Load determined per Columns B or C to Table 220.55

Applying Column B

Demand Percent (1 appliance) = .80
(8.1kW) 8100W x .80 = 6480W

Applying Column C

Maximum Demand (1 appliance)
6480W = 8000W

Use smallest demand load of Columns B (6480W) and C (8000W)

RANGE/COOKING APPLIANCES

6.	LINE LOAD	NEUTRAL LOAD (NEC 220.61(B)(1)
	6480VA	4536VA (6480VA x .70)

7. Heating and Air-Conditioning Equipment <NEC 220.50, 220.51, 220.60, Table 430.248 and 440.6(A)>

Heating Load (larger)

13.3kW (13,300VA)

AC Load

2743VA

Electric Heat   Blower
Line Load = 13,300VA + 785VA = 14,085VA

HEATING and AC

7.	LINE LOAD	NEUTRAL LOAD
	14,085VA	0

TOTAL DEMAND LOAD (LINE and NEUTRAL) (Add lines 1. − 8.) Highlights identify loads affected by re-calculations.

	LINE LOAD	NEUTRAL LOAD	
1. - 3.	3682.5VA	3682.5VA	
4.	6225.0VA	2850.0VA	
5.	0	0	
6.	6480.0VA	4536.0VA	
7.	14,085.0VA	0	
8.	150.0VA	150.0VA	(L.M. - 25% of 600VA disposal)
	30,622.5VA	11,218.5VA	

DWELLING'S OPERATING LINE VOLTAGE - 240V
(Given operating voltage or as determined per test examination)

CALCULATE MINIMUM LINE and NEUTRAL LOADS
(Divide Total Demand Load [VA] by operating line voltage [V])

LINE LOAD = 30,622.5VA / 240V = 127.59A
NEUTRAL LOAD = 11,218.5VA / 240V = 46.74A

SIZE SERVICE (Size of service based on the calculated **LINE LOAD**)

SIZE SERVICE REQUIRED (minimum) 150A
(Single rating - Main overcurrent device **150A**)

SIZING FEEDER/SERVICE CONDUCTORS

Using Table 310.15(B)(7), as a minimum use 1 AWG copper conductors (75°C) which has an ampacity of 130A as the service conductors.

FEEDER/SERVICE CONDUCTORS 1 AWG copper conductors

SIZING NEUTRAL CONDUCTOR(S)

Using Table 310.15(B)(16), as a minimum, use an 8 AWG copper conductor (75°C) which has an ampacity of 50A, as the neutral conductor. However, the use of an 8 AWG copper conductor as the neutral is not in compliance with NEC 250.24(C)(1) and Table 250.66. Therefore, as a minimum a 6 AWG copper conductor must be used instead as the neutral conductor.

NEUTRAL CONDUCTOR(S) ~~8~~ 6 AWG copper conductor

900SF unit - 240/120V 1-phase (900SF unit load calculation covers pages 163 – 165)

4. Appliance Loads (Fastened-In-Place) <NEC 220.53>

120V Appliances	VA Rating
1. Dishwasher	900
2. Disposal	600
3. Microwave	1200
4. Heat-Vent-Light	1600
	4300 (Total 120V Appliances)

240V Appliances	VA Rating
1. Water Heater	4500 (Total 4500V Appliance)
	4500

APPLIANCES TOTAL	=	8800VA

APPLY DEMAND FACTOR (Applicable, when number of above appliances exceeds four (4) or more.)

(Appliances Total) 8800VA x .75 = 6600VA

APPLIANCE LOADS

4.   <u>LINE LOAD</u>   <u>NEUTRAL LOAD (Refer to condition)</u>

        6600VA   3225VA

<u>Condition</u>
Because the water heater is a 240 volts line-to-line load it will not contribute to the neutral load as the other appliances and therefore must not be included. As a result, the NEUTRAL LOAD is equal to

$$8800VA - 4500VA \times .75 = \underline{3225VA}$$

5. Clothes Dryer (240/120V-1ϕ) <NEC and Table 220.54>

CLOTHES DRYER

5.   <u>LINE LOAD</u>   <u>NEUTRAL LOAD [NEC 220.61(B)(1)]</u>

        5000VA   3500VA (5000VA x .70)

6. Electric Ranges and Other Cooking Appliances (240/120V-1ϕ) <NEC and Table 220.55>

Maximum Demand Load determined per Columns C to Table 220.55.

        Applying Column C

                Maximum Demand (1 appliance)
                (10.4kW) 10,400VA = 8000W

RANGE/COOKING APPLIANCES

6.   <u>LINE LOAD</u>   <u>NEUTRAL LOAD [NEC 220.61(B)(1)]</u>

        8000VA   5600VA (8000VA x .70)

7. Heating and Air-Conditioning Equipment <NEC 220.50, 220.51, 220.60, Table 430.248 and 440.6(A)>

Heating Load - 16kW (16,000VA) (larger)

AC Load - 3429VA

                Electric Heat   Blower
Line Load = <u>16,000VA</u> + <u>884VA</u> = <u>16,884VA</u>

HEATING and AC

7.	LINE LOAD	NEUTRAL LOAD
	16,884VA	0

TOTAL DEMAND LOAD (LINE and **NEUTRAL)** (Add lines 1. – 8.) Highlights identify loads affected by re-calculations.

	LINE LOAD	NEUTRAL LOAD	
1. - 3.	4470VA	4470VA	
4.	6600VA	3225VA	
5.	5000VA	3500VA	
6.	8000VA	5600VA	
7.	16,884VA	0	
8.	150VA	150VA	(L.M. - 25% of 600VA disposal)
	41,104VA	16,945VA	

DWELLING'S OPERATING LINE VOLTAGE - 240V
(Given operating voltage or as determined per test examination)

CALCULATE MINIMUM LINE and **NEUTRAL LOADS**
(Divide Total Demand Load **[VA]** by operating line voltage **[V]**)

LINE LOAD = 41,104VA / 240V = 171.27A
NEUTRAL LOAD = 16,945VA / 240V = 70.60A

SIZE SERVICE (Size of service based on the calculated **LINE LOAD**)

SIZE SERVICE REQUIRED (minimum) <u>175A</u>
(Single rating - Main overcurrent device **175A**)

SIZING FEEDER/SERVICE CONDUCTORS

Using Table 310.15(B)(7), as a minimum use 1/0 AWG copper conductors (75°C) which has an ampacity of 150A as the service conductors.

FEEDER/SERVICE CONDUCTORS <u>1/0 AWG copper conductors</u>

SIZING NEUTRAL CONDUCTOR(S)

Using Table 310.15(B)(16), as a minimum, use a 4 AWG copper conductor (75°C) which has an ampacity of 85A, as the neutral conductor. The use of a 4 AWG copper conductor as the neutral is in compliance with NEC 250.24(C)(1) and Table 250.66.

NEUTRAL CONDUCTOR(S) <u>4 AWG copper conductor</u>

1200SF unit - 240/120V 1-phase (1200SF unit load calculation covers pages 166 – 169)

4. Appliance Loads (Fastened-In-Place) <NEC 220.53>

120V Appliances	**VA Rating**
1. Dishwasher	1250
2. Disposal	900
3. Trash Compactor	780
4. Microwave	1600
5. Heat-Vent-Light	<u>1600</u>
	6130 (Total 120V Appliances)

240V Appliances	**VA Rating**
1. Water Heater	<u>4500</u> (Total 240V Appliance)
	4500

APPLIANCES TOTAL = 10,630VA

APPLY DEMAND FACTOR (Applicable, when number of above appliances exceeds four (4) or more.)

(Appliances Total) <u>10,630VA</u> x .75 = 7972.5VA

APPLIANCE LOADS

4.	LINE LOAD	NEUTRAL LOAD (Refer to condition)
	7972.5VA	4597.5VA

Condition
Because the water heater is a 240 volts line-to-line load it will not contribute to the neutral load as the other appliances and therefore must not be included. As a result, the NEUTRAL LOAD is equal to

10,630VA - 4500VA x .75 = 4597.5VA

5. Clothes Dryer (240/120V-1ϕ) <NEC and Table 220.54>

CLOTHES DRYER

5.	LINE LOAD	NEUTRAL LOAD [NEC 220.61(B)(1)]
	5000VA	3500VA (5000VA x .70)

6. **Electric Ranges and Other Cooking Appliances** (240/120V-1φ) <NEC and Table 220.55>

Minimum Demand Load determined per Column B and Column C.

Applying Column B

Demand Percent (2 appliances) = .65
(7.18kW) 7180W + (8.4kW) 8400W x .65 = 10.127kW (10,127VA)

Applying Column C

Maximum Demand (2 appliances)
11kW (11,0000W)

Use the smallest demand load of Column B and Column C.

REMINDER: Although it may appear that the initial provision of **Note 4** to Table 220.55 could have been applied toward the use of the two cooking appliances, it cannot. Because both cooking appliances are used toward calculating the service load for this occupancy such provision is not applicable. The provisions of **Note 4** to Table 220.55 are only applicable for branch-circuit loads and not for service and feeder load calculations. If the provision of **Note 4** to Table 220.55 could be applied in treating the two cooking appliances as one range, the provisions of **Note 1** to Table 220.55 would be required to obtain a demand load; where in this situation it would produce a demand load smaller than 10.13kW.

NEUTRAL LOAD

Based on the exclusive use of the WMO per Column B the neutral load is,

(7.18kW) 7180W x .80 = 5744W

RANGE/COOKING APPLIANCES

6.   LINE LOAD   NEUTRAL LOAD [NEC 220.61(B)(1)]

10,130VA   4021VA (WMO) (5744VA x .70)

7. **Heating and Air-Conditioning Equipment** <NEC 220.50, 220.51, 220.60, Table 430.248 and 440.6(A)>

Heating Load

20kW (20,000VA)

AC Load

      4114VA

Heating load is larger.

            Electric Heat   Blower
Line Load = 20,000VA + 992VA = 20,992VA

HEATING and AC

7.	LINE LOAD	NEUTRAL LOAD
	20,992VA	0

TOTAL DEMAND LOAD (LINE and **NEUTRAL)** (Add lines 1. – 8.) Highlights identify loads affected by re-calculations.

	LINE LOAD	NEUTRAL LOAD
1. - 3.	4785.0VA	4785.0VA
4.	7972.5VA	4597.5VA
5.	5000.0VA	3500.0VA
6.	10,130.0VA	4021.0VA
7.	20,992.0VA	0
8.	225.0VA	225.0VA
	49,104.5VA	17,128.5VA

DWELLING'S OPERATING LINE VOLTAGE - 240V
(Given operating voltage or as determined per test examination)

CALCULATE MINIMUM LINE and **NEUTRAL LOADS**
(Divide Total Demand Load **[VA]** by operating line voltage **[V]**)

LINE LOAD = 49,104.5VA / 240V = 204.60A
NEUTRAL LOAD = 17,128.5VA / 240V = 71.37A

SIZE SERVICE (Size of service based on the calculated **LINE LOAD**)

SIZE SERVICE REQUIRED (minimum) **225A**
(Single rating - Main overcurrent device **225A**)

SIZING FEEDER/SERVICE CONDUCTORS

Using Table 310.15(B)(7), as a minimum use 3/0 AWG copper conductors (75°C) which has an ampacity of 200A as the service conductors.

FEEDER/SERVICE CONDUCTORS 3/0 AWG copper conductors

SIZING NEUTRAL CONDUCTOR(S)

Using Table 310.15(B)(16), as a minimum, use a 4 AWG copper conductor (75°C) which has an ampacity of 85A, as the neutral conductor. The use of a 4 AWG copper conductor as the neutral is in compliance with NEC 250.24(C)(1) and Table 250.66.

NEUTRAL CONDUCTOR(S) 4 AWG copper conductor

1400SF unit - 240/120V 1-phase (1400SF unit load calculation covers pages 169 – 172)

4. Appliance Loads (Fastened-In-Place) <NEC 220.53>

120V Appliances	VA Rating
1. Dishwasher	1250
2. Disposal	900
3. Trash Compactor	780
4. Microwave	1800
5. Heat-Vent-Light	1750
	6480 (Total 120V Appliances)

240V Appliances	VA Rating
1. Water Heater	4500 (Total 240V Appliance)
	4500

APPLIANCES TOTAL   =   10,980VA

APPLY DEMAND FACTOR (Applicable, when number of above appliances exceeds four (4) or more.)

(Appliances Total) 10,980VA x .75 = 8235VA

APPLIANCE LOADS

4.	LINE LOAD	NEUTRAL LOAD (Refer to condition)
	8235VA	4860VA

Condition

Because the water heater is a 240 volts line-to-line load it will not contribute to the neutral load as the other appliances and therefore must not be included. As a result, the NEUTRAL LOAD is equal to

10,980VA - 4500VA x .75 = 4860VA

5. Clothes Dryer (240/120V-1φ) <NEC and Table 220.54>

CLOTHES DRYER

5.   <u>LINE LOAD</u>   <u>NEUTRAL LOAD [NEC 220.61(B)(1)]</u>

       5000VA   3500VA (5000VA x .70)

6. Electric Ranges and Other Cooking Appliances (240/120V-1φ) <NEC and Table 220.55>

Minimum Demand Load determined per Column B and Column C.

   Applying Column B – only 7.7kW WMO applicable; 8.9kW CT per Column C
      Considering both appliances - the only applicable method is per Column C.

   Applying Column C
      Maximum Demand (2 appliances)
      11kW(11,0000W)

Use the demand load of Column C.

REMINDER: Although it may appear that the initial provision of **Note 4** to Table 220.55 could have been applied toward the use of the two cooking appliances, it cannot. Because both cooking appliances are used toward calculating the service load for this occupancy such provision is not applicable. The provisions of **Note 4** to Table 220.55 are only applicable for branch-circuit loads and not for service and feeder load calculations. If the provision of **Note 4** to Table 220.55 could be applied in treating the two cooking appliances as one range, the provisions of **Note 1** to Table 220.55 would be required to obtain a demand load; where in this situation it would produce a demand load smaller than 11kW.

Based on the exclusive use of the WMO per Column B the neutral load is,

$$(7.7kW)\ 7700W\ x\ .80 = 6160VA$$

RANGE/COOKING APPLIANCES

6.   <u>LINE LOAD</u>   <u>NEUTRAL LOAD [NEC 220.61(B)(1)]</u>

       11,000VA   4312VA (6160VA x .70)

7. Heating and Air-Conditioning Equipment <NEC 220.50, 220.51, 220.60, Table 430.248 and 440.6(A)>

Heating Load - 24kW(24,000VA) (larger)

AC Load - 4800VA

Electric Heat   Blower
Line Load = 24,000VA + 1065VA = 25,065VA

HEATING and AC

7.   LINE LOAD   NEUTRAL LOAD

       25,065VA   0

TOTAL DEMAND LOAD (LINE and **NEUTRAL)** (Add lines 1. – 8.) Highlights identify loads affected by re-calculations.

	LINE LOAD	NEUTRAL LOAD
1. - 3.	4995VA	4995VA
4.	8235VA	4860VA
5.	5000VA	3500VA
6.	11,000VA	4312VA
7.	25,065VA	0
8.	225VA	225VA
	54,520VA	17,892VA

DWELLING'S OPERATING LINE VOLTAGE - 240V
(Given operating voltage or as determined per test examination)

CALCULATE MINIMUM LINE and **NEUTRAL LOADS**
(Divide Total Demand Load **[VA]** by operating line voltage **[V]**)

LINE LOAD = 54,520VA / 240V = 227.17A
NEUTRAL LOAD = 17,892VA / 240V = 74.6A

SIZE SERVICE (Size of service based on the calculated **LINE LOAD**)

SIZE SERVICE REQUIRED (minimum) 250A
(Single rating - Main overcurrent device **250A**)

SIZING FEEDER/SERVICE CONDUCTORS

Using Table 310.15(B)(7), as a minimum use 4/0 AWG copper conductors (75°C) which has an ampacity of 230A as the service conductors.

FEEDER/SERVICE CONDUCTORS 4/0 AWG copper conductors

SIZING NEUTRAL CONDUCTOR(S)

Using Table 310.15(B)(16), as a minimum, use a 2 AWG copper conductor (75°C) which has an ampacity of 115A, as the neutral conductor. The use of a 2 AWG copper conductor as the neutral is in compliance with NEC 250.24(C)(1) and Table 250.66.

NEUTRAL CONDUCTOR(S) 2 AWG copper conductor

COMPARISION OF RESULTS

208/120V-3ϕ vs 240/120V-1ϕ

LOAD	Line (L) Amps L. Conductors	Neutral (N) Amps N. Conductors	Size Service	Line Amps L. Conductors	Neutral Amps N. Conductors	Size Service
House	593.85A NA	70.29A NA	600A	1070.62A NA	105.5A NA	1200A
54 units	3805.4A NA	1018.43A NA	4000A	6785.37A NA	1207.52A NA	7000A
15 units 650SF	848.68A 3-400 kcmil	225A 3-1/0 AWG	1000A	1511.41A 4-600 kcmil	327.86A 4-1/0 AWG	1600A
17 units 900SF	1245.38A 4-600 kcmil	328.88A 4-1/0 AWG	1600A	2128.24A 4-1500 kcmil	486.77A 4-1/0 AWG	2500A
12 units 1200SF	1090.72A 3-600 kcmil	292.42A 3-1/0 AWG	1200A	1876.54A 4-900 kcmil	437.96A 4-1/0 AWG	2000A
10 units 1400SF	1051.29A 3-600 kcmil	274.31A 3-1/0 AWG	1200A	1773.44A 4-900 kcmil	398.99A 4-1/0 AWG	2000A

208/120V-1ϕ vs 240/120V-1ϕ

LOAD	Line (L) Amps L. Conductors	Neutral (N) Amps N. Conductors	Size Service	Line Amps L. Conductors	Neutral Amps N. Conductors	Size Service
650SF Single	121.41A 2 AWG	54.48A 6 AWG	125A	127.59A 1 AWG	46.74A 6 AWG	150A
900SF Single	167.98A 2/0 AWG	91.03A 3 AWG	175A	171.27A 1/0 AWG	70.60A 4 AWG	175A
1200SF Single	197.44A 3/0 AWG	90.99A 3 AWG	200A	204.60A 3/0 AWG	71.37A 4 AWG	225A
1400SF Single	217.1A 3/0 AWG	94.81A 3 AWG	225A	227.17A 4/0 AWG	74.6A 4 AWG	250A

NONDWELLING BUILDINGS (COMMERCIAL and INDUSTRIAL)

Before getting started with commercial and industrial calculations let's take a look at the different types of voltage combinations one could encounter while reviewing this section. Whether it involves preparing for an examination or performing such load calculations in a real world situation, electrical calculations in general has always proven to be quite difficult for most. Because of the need to operate certain loads and equipment at different voltages and phase combinations, it is important to discuss this very practical realization before taking on such challenges that may lie ahead.

When performing load calculations or designing electrical systems single and three-phase voltage combinations are for certain. In a single-phase 3-wire system; 120V, 208V, 240V, 208/120V and 240/120V are sources of voltage that some or perhaps most electrical professional are familiar with. These voltage sources normally are not nearly as confusing or intimidating as 3- or 4-wire, 3-phase systems. With a three-phase system both single- and three-phase loads can be supplied from the same source. Considering the most common three-phase voltages be it, 208/120V, 240/120V or 480/277V 4-wire systems there are at least three voltage combinations that can be derived from each system. This includes single-phase line-to-neutral and line-to-line loads and three-phase line to line loads. From 208/120V and 240/120V 4-wire, three-phase systems single-phase 3-wire systems can be supplied. When calculating voltage and phase combinations above these commonly used sources (that is, 500V to 600V, 2400V, 4160V, 7200V, 13,800V, etc.) the same approach is applied. The most important thing to remember when performing load calculations is, regardless of voltage combinations or phases always use the highest voltage in the denominator of the given formula which in essence will include all lower voltage combinations.

Standard Load Calculations for Nondwelling Buildings (Commercial and Industrial) (89. - 103.) [15]

Refer to (**Worksheet G** - Volume 4) STANDARD LOAD CALCULATION FOR NONDWELLING BUILDINGS (COMMERCIAL and INDUSTRIAL) for related questions.

Assembly Hall, Auditorium, Corridor

89. A major university undergoing new construction and renovation will include the addition of two assembly halls, an auditorium and adjoining corridors. The dimensions of these new occupancies are as follows:

Assembly 1 - 87' x 95'
Assembly 2 - 70' x 90'
Auditorium - 235' x 310'
Corridor 1 - 7.5' x 50'
Corridor 2 - 8' x 68'
Corridor 3 - 9' x 92'
Corridor 4 - 9.5' x 137'

Determine each individual general lighting load and the total general lighting load for these occupancies based on a continuous service demand.

GENERAL LIGHTIN LOAD

General Lighting Load <NEC References - 220.12 and Table 220.12>

Occupancies	Individual general lighting loads
Assembly 1 - 87 ft. x 95 ft. x 1VA/sft. =	8265.00VA
Assembly 2 - 70 ft. x 90 ft. x 1VA/sft. =	6300.00VA
Auditorium - 235 ft. x 310 ft. x 1VA/sft. =	72,850.00VA
Corridor 1 - 7.5 ft. x 50 ft. x ½VA/sft. =	187.50VA
Corridor 2 - 8 ft. x 68 ft. x ½VA/sft. =	272.00VA
Corridor 3 - 9 ft. x 92 ft. x ½VA/sft. =	414.00VA
Corridor 4 - 9.5 ft. x 137 ft. x ½VA/sft. =	650.75VA
	Total 88,939.25VA

Applying the provision of NEC 230.42(A)(1) the total general lighting load for these occupancies is,

$$88,939.25VA \times 1.25 = 111,174.1VA$$

Banks [2]

90. A 60' x 50' single-story credit union will be supplied by a 240/120V, 1-phase, 3W service. The following loads are included:

120V	240V
Actual interior lighting - 16,000VA	Water Heater - 6.5kW
Receptacles - unknown	Electric Heating Unit No. 1 - 10kW
Lighting Track - 125ft.	Electric Heating Unit No. 2 - 20kW
Sign - 9A	Blower Motor No. 1 - 5.3A
Parking Lights - 63A	Blower Motor No. 2 - 7.5A
	AC Compressor Motor No. 1 - 17.83A
	AC Compressor Motor No. 2 - 25.27A
	AC Fan Motor Nos. 1 and 2 - 2.6A

What size service is required to serve the building? At 75°C, what size service, neutral and grounding electrode copper conductors are required to support the building's electrical system?

1. GENERAL LIGHTING or ACTUAL LIGHTING LOADS

General Lighting Load <NEC References – 220.12 and Table 220.12>

$$60' \times 50' \times 3.5VA = 10,500VA$$

A. General Lighting Load = ~~10,500VA~~ <Omit- General Lighting Load - Actual Lighting Load Larger>

Actual Lighting Load

B. Actual Lighting Load = 16,000VA

$$16,000VA \times 1.25 = 20,000VA$$

1.

LINE LOAD	NEUTRAL LOAD
20,000VA	16,000VA

2. OTHER LIGHTING LOADS

A. Sign/Outline (S/O) Lighting <NEC References - 220.12(F) and 600.5(A)>

Sign - 120V x 9A = ~~1080VA~~

Total (Sign/Outline Lighting) = 1200VA (required minimum)

B. Outside Lighting <NEC Reference - 220.18(B)>

120V x 63A = 7560VA

Total (Outside Lighting) = 7560VA

C. NA

D. Track Lighting <NEC Reference - 220.43(B)> (Voltage rating -120V)

$$(125' \div 2') \times 150VA = 9375VA$$

E. NA

F. Other Lighting (LINE and NEUTRAL) Loads Total [Add lines (A.) - (E.) where applicable]

Other Lighting Lights Total = 18,135VA

$$TOTAL = 18,135VA \times 1.25 = 22,688.75VA$$

2.

LINE LOAD	NEUTRAL LOAD
22,668.75VA	18,135VA

3. RECEPTACLE LOADS

A. NA

B. Non-continuous duty (Bank) <NEC 220.14(K)> Because the number of receptacles are unknown NEC 220.14(K)(1)[220.14(I)] cannot be applied. Only NEC 220.14(K)(2) is applicable.

    (1) NA

    (2) 60' x 50' x 1VA/SF = 3000VA

                  Total Receptacle loads = 3000VA

C. NA

D. Total Receptacle (LINE and NEUTRAL) Load
   LINE = 3000VA
   NEUTRAL = 3000VA

3.	LINE LOAD	NEUTRAL LOAD
	3000VA	3000VA

4. KITCHEN EQUIPMENT - NA

5. SPECIFIC LOADS

Type Load	Calculation
240V	
Water Heater	6500W

5.	LINE LOAD	NEUTRAL LOAD
	6500VA	0

6. MOTOR LOADS - NA

7. MEDICAL EQUIPMENT - NA

8. INDUSTRIAL EQUIPMENT - NA

9. HEATING and AIR-CONDITIONING (AC) EQUIPMENT <NEC References 220.50, 220.51, 220.60, 430.6(A)(1) and 440.6(A)>

ELECTRICAL HEATING UNITS

Heating Unit	Calculation
No. 1 - 10kW	10,000W
No. 2 - 20kW	20,000W

TOTAL HEAT = 30,000W(VA)

AIR-CONDITIONING (AC) UNITS

AC Unit	Calculation
No. 1	240V x 17.83A + 240V x 2.6A = 4903.2VA
No. 2	240V x 25.27A + 240V x 2.6A = 6688.8VA

TOTAL AC = 11,592.0VA

Line Load = 30,000VA   +   (3072VA)   = 33,072VA
               (Largest Load)   (Blowers)

Blowers - 240V x (5.3A + 7.5A) = 3072VA

9.	LINE LOAD	NETRAL LOAD
	33,072VA	0

10. LARGEST MOTOR

AC Compressor Motor No. 2 - 240V x 25.27A x .25(25 percent) = 1516.2VA

10.	LINE LOAD	NEUTRAL LOAD
	1516.2VA	0

TOTAL DEMAND LOAD (LINE and NEUTRAL) (List each computed line and neutral loads below and total lines 1. - 10.)

	LINE LOAD	NEUTRAL LOAD
1. General Lighting	20,000.00VA	16,000VA
2. Other Lighting Loads	22,668.75VA	18,135VA
3. Receptacle Loads	3000.00VA	3000VA
4. Kitchen Equipment	--	--

5.	Specific Loads	6500.00VA	0
6.	Motor Loads	--	--
7.	Medical Equipment	--	--
8.	Industrial Equipment	--	--
9.	Heating and AC Equipment	33,072.00VA	0
10.	Largest Motor	1516.20VA	0

TOTAL = 86,756.95VA 37,135VA

OCCUPANCY'S OPERATING LINE VOLTAGE - 240V
(Given operating voltage or as determined per test examination)

CALCULATE MINIMUM LINE and NEUTRAL LOADS
(Divide Total Demand Load **[VA]** by operating line voltage **[V]**)

LINE LOAD = 86,756.95VA / 240V = 361.49A
NEUTRAL LOAD = 37,135VA / 240V = 154.73A

SIZE SERVICE (Size of service based on the calculated **LINE LOAD**)

SIZE SERVICE REQUIRED (minimum) 400A
(Single rating or combination - Main overcurrent device(s) to total **400A**)

SIZING FEEDER/SERVICE CONDUCTORS <NEC References - NEC 215.2(A), 230.42(A), 240.4(B) & (C), 310.10(H), 310.15(B)(2) & (3) and Table 310.15(B)(16)> (Based on the calculated LINE LOAD and NEC References.)

Per Table 310.15(B)(16) at 75°C the minimum size copper conductors required to supply a 400A service are 500 kcmil which have a rated ampacity of 380 amps. Although the ampacity of the conductors is less than the 400A service they will supply, the use of these conductors are permitted per NEC 240.4(B).

FEEDER/SERVICE CONDUCTORS 500 kcmil copper

SIZING NEUTRAL CONDUCTOR <NEC References - 215.2(A)(2), 220.61, 230.42(C), 250.24(C), 310.10(H), 310.15(B)(2), (3) & (5) and Table 310.15(B)(16) > (Based on the calculated NEUTRAL LOAD and NEC References.)

Per Table 310.15(B)(16) at 75°C the minimum size copper conductor required to serve as the neutral conductor is a 2/0 AWG which has a rated ampacity of 175 amps.

NEUTRAL CONDUCTOR(S) 2/0 AWG copper

SIZING GROUNDING ELECTRODE CONDUCTOR <NEC References - 250.24(C), 250.66 & Table 250.66 and Table 8 of Chapter 9>

Per Table 250.66 based on the use of 500 kcmil copper service conductors the grounding electrode conductor must be a 1/0 AWG copper which is smaller than the 2/0 AWG copper neutral conductor.

GROUNDING ELECTRODE CONDUCTOR 1/0 AWG copper

91. An 18,000SF four story central bank building will be supplied by a 480/277V, 3-phase, 4W service. The following loads and other occupancy are included:

120V
Receptacles - 183 (non-continuous)
Receptacles - 27 (continuous)
Plugmolds - 70ft. (heavy duty)
Plugmolds - 70ft. (light duty)
Soda Machines (2) - 1480VA
Water fountains (4) - 6.5A
Desktop computers (22) - 1.6A
Laser printers (10) - 1.3A
Exhaust fans (6) - ½ HP
Microwave ovens (2) - 1600W
Coffee makers (3) - 1540W

277V
Outside lighting (15) - 165VA/fixture
Show Window - 72ft.
Signs (2) - 2340VA (fluorescent)
Track Lighting - 245ft.

208V-1φ
Water Heaters (2) - 8kW and 5kW

208V-3φ
Main frame computers (2) - 17,500VA
  (continuous-duty)
Copiers (commercial grade) (4) - 15.7A
  (continuous-duty)

480V-1φ
Outside lighting (15) - .89A

480V-3φ
Heating units (4) - 50kW
Blower motors (4) - 8.3A
Air Handlers (4) - 12.3A
AC units (4) - 44,000VA
AC fan motors (8) - 3.6A
Fire pump motor - 50HP
Jockey pump motor - 3HP
Elevator motors (2) - 60HP
Intermittent duty/continuous - NPC - 72A
Elevator motors (2) - 60HP
Intermittent duty/15-minute - NPC - 68A

Other Occupancy
Stairways - 700SF

What size service is required to serve the bank? Using THWN copper conductors, what size service, neutral and grounding electrode conductors are required to supply the service? What size parallel conductors are needed, if using 3 conductors per line? Assume all general lighting loads are 277V where 80 percent are non-linear.

1. GENERAL LIGHTING or ACTUAL LIGHTING LOADS

General Lighting Load <NEC References - 220.12 and Table 220.12>

$$\begin{array}{l} \text{Bank - 18,000SF} \quad \text{x 3.5VA} = 63{,}000\text{VA} \\ \text{Stairway - 700SF} \quad \text{x ½VA} = \underline{\quad 350\text{VA}} \\ \hspace{6cm} 63{,}350\text{VA} \end{array}$$

A. General Lighting Load - 63,350VA x 1.25 = 79,187.5VA

Actual Lighting Load

B. Actual Lighting Load - NA

1.	LINE LOAD	NEUTRAL LOAD	
		Permitted Reduction	Prohibited Reduction
	79,187.5VA	0	63,350VA

2. OTHER LIGHTING LOADS

A. Sign/Outline (S/O) Lighting <NEC References - 220.12(F) and 600.5(A)>

Signs - (2) (2340VA @ 277V*) - 2340VA x 2 = 4680VA

Total (Sign/Outline Lighting) = 4680VA

B. Outside Lighting <NEC Reference - 220.18(B)>

277V*   - 165VA x 15   =   2475.0VA
480V (1φ)** - 480V x .89A x 15 =   6408.0VA

Total (Outside Lighting) =   8883.0VA

C. Show-Window Lighting <NEC Reference - 220.43(A)> (Voltage rating - 277V*)

72' x 200VA = 14,400VA

D. Track Lighting <NEC Reference - 220.43(B)> (Voltage rating - 277V*)

(245' ÷ 2') x 150VA = 18,375VA

E. Miscellaneous NA

F. Other Lighting (LINE and NEUTRAL) Loads Total [Add lines (A.) - (E.) where applicable]

Other Lighting Lights Total = 46,338.0VA

TOTAL = 46,338.0VA x 1.25 = 57,922.5VA

2. <u>LINE LOAD</u>

<u>NEUTRAL LOAD</u>

Permitted Reduction	Prohibited* Reduction

57,922.5VA   0   39,930VA**

**480V (1φ) Fixtures (not included, line-to-line)

3. RECEPTACLE LOADS

A. Non-continuous duty <NEC References - 220.14(H) and (I), 220.44 and Table 220.44>

(1) Fixed Multioutlet Assemblies <NEC Reference - 220.14(H)>

Non-simultaneous use (Plugmolds-light duty) - (70' ÷ 5') x 180VA = 2520VA

Simultaneous use (Plugmolds-heavy duty) - 70' x 180VA = 12,600VA

(2) General Purpose Receptacles* and Fixed Multioutlet Assemblies <NEC References – 220.14(I)> (120V)

0* + 2520VA + 12,600VA + 32,940VA** = 48,060VA

**Since the results of B. (1) is larger than B. (2) below, this method can be omitted thus allowing the receptacles [183@32,940VA] to be combined with the plugmolds of step 3.A(1) to obtain a smaller receptacle demand between the two non-continuous receptacle loads opposed to gathering a larger demand using two individual calculations.

APPLY DEMAND FACTORS
a. First 10,000VA (10kVA) [@100%] = 10,000VA
b. Remainder - 38,060VA x .50   = <u>19,030VA</u>
                          Total = 29,030VA [Receptacle load per 3.A(1) & B.]

B. Non-continuous duty (Bank) <NEC 220.14(K)> (120V)

(1) 183 x 180VA = ~~32,940VA~~ (Larger based on results of demand factors - 21,470VA)

APPLY DEMAND FACTORS

   a. First 10,000VA (10kVA) [@100%] = 10,000VA

   b. Remainder - 22,949VA x .50   = <u>11,470VA</u>

                            Total   = 21,470VA

  (2) 18,000SF x 1VA/SF = 9300VA

C. Continuous duty <NEC References - 230.42(A)> (120V)

   27 x 180VA = 4860VA x 1.25 = 6075VA

D. Total Receptacle (LINE and NEUTRAL) Load
   LINE = 29,030VA + 6075VA = 35,105VA
   NEUTRAL = 29,030VA + 4860VA = 33,890VA

3.   <u>LINE LOAD</u>   <u>NEUTRAL LOAD</u>

	Permitted Reduction	Prohibited Reduction
35,105VA	33,890VA	0

4. KITCHEN EQUIPMENT - NA

5. SPECIFIC LOADS

Type Load	Calculation		
120V			
Coffee makers	1540W x 3	=	4,620.00W(VA)
Desktop computers*	120V x 1.6A x 22	=	4,224.00VA
Laser printers*	120V x 1.3A x 10	=	1,560.00VA
Microwave ovens	1600W x 2	=	<u>3,200.00W(VA)</u>
			13,604.00VA
208V(1ϕ)			
Water Heaters	8000W + 5000W	=	13,000.00W(VA)
208V(3ϕ)			
Main frame computers**	17,500VA x 2 x 1.25	=	43,750.00VA
Copiers	22,624.08VA*** x 1.25	=	<u>28,280.10VA</u>
			72,030.10VA

  ** See question No. 1. of Article 645
*** 208V x 15.7A x 1.732 x 4 = 22,624.08VA   TOTAL = 98,634.10VA

5.

LINE LOAD	NEUTRAL LOAD	
	Permitted Reduction	Prohibited* Reduction
98,634.10VA	7820VA	5784VA

6. MOTOR LOADS <NEC Reference - 220.50 >

A. Continuous Duty <NEC References - Tables 430.247 - 430.250>

Motor Load	Calculation		
120V			
½HP exhaust fans	120V x 9.8A* x 6	=	7056.00VA
208V-3φ			
3HP Jockey pump	208V x 10.6A* x 1.732	=	3818.71VA
50HP Fire pump	208V x 143A* x 1.732	=	51,516.61VA
*Tables 430.248-(1φ) and 430.250-(3φ)			55,335.32VA

B. Other Than Continuous Duty

120V			
Soda machines	1480VA x 2	=	2960.00VA
Water fountains	120V x 6.5A x 4	=	3120.00VA
			6080.00VA

C. NA

D. Elevators (4) <NEC Reference - Table 430.22(E)>

60HP motors-Intermit. dty/cont. (2) - 480V x 72A x 1.732 x 2 x 1.40 = 167,602.18VA
60HP motors-Intermit. dty/15min. (2) - 480V x 68A x 1.732 x 2 x .85 = 96,105.22VA
~~263,707.40VA~~

263,707.40VA x .85** = 224,151.29VA (Demand Load)
**Apply Demand Factor (Table 620.14) for Elevators (4 @ .85) (assumed - not under constant load)

E. Total Motor Loads - [LINE LOAD - Add lines (A.), (B.) and (D.)] - [NEUTRAL LOAD - Total motor loads with neutral connections (120V)]

TOTAL = 292,622.61VA

6.

LINE LOAD	NEUTRAL LOAD	
	Permitted Reduction	Prohibited Reduction
292,622.61VA	13,136VA	0

7. MEDICAL EQUIPMENT - NA

8. INDUSTRIAL EQUIPMENT - NA

9. HEATING and AIR-CONDITIONING (AC) EQUIPMENT <NEC References - 220.50, 220.60, 430.6(A)(1) and 440.6(A)>

ELECTRICAL HEATING UNITS

Heating Unit	Calculation
50kW (4)	50,000W x 4 = 200,000(W)VA
	TOTAL HEAT = 200,000VA

AIR-CONDITIONING (AC) UNITS

AC Unit	Calculation	
44,000VA (4)	44,000VA x 4	= 176,000.00VA
Fan Motors (8 - 3.6A)	480V x 3.6A x 1.732 x 8 =	23,943.17VA
	TOTAL AC	= 199,943.17VA

$$\text{Line Load} = \underset{\text{(Larger Load)}}{\overset{\text{(Heat)}}{200{,}000\text{VA}}} + \underset{\text{(Air Handlers/Blowers)}}{(68{,}504.06\text{VA*})} = 268{,}504.06\text{VA}$$

Air Handlers - 480V x 12.3A x 1.732 x 4 = 40,902.91VA
Blowers -   480V x 8.3A x 1.732 x 4 = 27,601.15VA
                                   *68,504.06VA

9.	LINE LOAD	NEUTRAL LOAD	
		Permitted Reduction	Prohibited Reduction
	268,504.06VA	0	0

10. LARGEST MOTOR

Elevator (motor with highest current per NEC 430.17 and 440.7)

480V x 77A x 1.732 x .25 (25 percent) = 16,003.68VA

10.	LINE LOAD	NEUTRAL LOAD	
		Permitted Reduction	Prohibited Reduction
	268,504.06VA	0	0

TOTAL DEMAND LOAD (LINE and NEUTRAL) (List each computed line and neutral loads below and total lines 1. - 10.)

		LINE LOAD	NEUTRAL LOAD	
			Permitted Reduction	Prohibited Reduction
1.	General Lighting	79,187.50VA	0	63,350VA
2.	Other Lighting Loads	57,922.50VA	0	39,930VA
3.	Receptacle Loads	35,105.00VA	33,890VA	0
4.	Kitchen Equipment	--	--	--
5.	Specific Loads	98,634.10VA	7,820VA	5,784VA
6.	Motor Loads	292,622.61VA	13,136VA	0
7.	Medical Equipment	--	--	--
8.	Industrial Equipment	--	--	--
9.	Heating and AC Equipment	268,504.06VA	0	0
10.	Largest Motor	16,003.68VA	0	0
	TOTAL =	847,979.45VA	55,846VA	109,064VA

OCCUPANCY'S OPERATING LINE VOLTAGE - 480V(3φ)
(Given operating voltage or as determined per test examination)

CALCULATE MINIMUM LINE AND NEUTRAL LOADS
(Divide Total Demand Load [VA] by operating line voltage [V])

LINE LOAD = 847,979.45VA / 480V x 1.732 = 1020A
NEUTRAL LOAD

  Permitted   =   55,846VA / 480V x 1.732 =   71.99A
  Prohibited   =   109,064VA / 480V x 1.732 = 131.19A

  Total Neutral Load = 67.17A + 131.19A = 198.36A

SIZE SERVICE (Size of service based on the calculated **LINE LOAD**)

SIZE SERVICE REQUIRED (minimum) 1200A
(Single rating or combination - Main overcurrent device(s) to total **1200A**)

SIZING FEEDER/SERVICE CONDUCTORS

In this problem three individual raceways will be used for this installation to accommodate 3 parallel conductors per line (L_1, L_2, and L_3). Considering the size service required, the ampacity of each individual parallel conductor can be determined. Therefore,

$$\frac{1200A}{3} = 400A \text{ (needed ampacity per Line conductors)}$$

Based on the calculated results each individual conductor must have an ampacity that is either equal to or exceeds 400 amps. Because THWN copper conductors are required per Table 310.15(B)(16) as a minimum, three 600 kcmil conductors which are individually rated for 420A appears to be the conductors of choice but first let's take a look at NEC 240.4(C).

NEC 240.4(C) requires the ampacity of conductors protected by an overcurrent device exceeding 800A to be equal to or greater than the rating of the overcurrent device. Being the case, a 600 kcmil copper conductor yields a total ampacity of 1260A (420A x 3) which is more than the minimum (1200A) rating of the required overcurrent device. The selected service conductors were based upon the use of a single 1200A main overcurrent device.

FEEDER/SERVICE CONDUCTORS 3-600 kcmil copper (per line)

SIZING NEUTRAL CONDUCTOR(S)

Based on the calculated neutral load and the use of parallel service conductors, the neutral conductors must also be installed in parallel per NEC 310.10(H)(1). Because the service conductors will consist of three parallel conductors per phase, the neutral conductors must also consist of three parallel conductors based on the calculated neutral load being divided by three which results to 67.73A (203.18A / 3). Per Table 310.15(B)(16), at 75°C, a 4 AWG copper conductor which has a rated ampacity of 85A will satisfy the calculated neutral load being distributed amongst three such conductors. However, because NEC 310.10(H)(1) only allows a 1/0 AWG conductor or larger to be installed in parallel, three 1/0 AWG copper conductors must be used instead as the parallel neutral conductors.

NEUTRAL CONDUCTOR(S) 3-1/0 AWG copper conductors

SIZING GROUNDING ELECTRODE CONDUCTOR

Based on the total cross-sectional area of the selected copper service conductors (1200 kcmil-[400 kcmil x 3]), per Table 250.66, a 3/0 AWG copper grounding electrode conductor must be used. Where the equivalent area for parallel copper service-entrance conductors exceeds 1100 kcmil, the grounding electrode conductor is not required to be larger than 3/0 AWG copper for any installation.

Although NEC 250.24(C)(1) requires the neutral (grounded) conductor to be either the same size or larger than the grounding electrode conductors, the total circular mils of the three 1/0 AWG neutral conductors (316,800 cmils [105,600 cmils x 3]) exceeds the circular mils of a 3/0 AWG conductor (167,800 cmils) per Table 8 of Chapter 9. As a result, this classifies the use of the three 1/0 AWG parallel neutral conductors as being larger than the 3/0 AWG grounding electrode conductor.

GROUNDING ELECTRODE CONDUCTOR 1-3/0 copper

Barber/Beauty Shop

92. Supplied by a 4-wire, 208/120V, 3-phase service, a 3965SF barber and beauty salon will consume the following loads:

(2) Washers - 1500VA, 120V
(2) Dryers - 5.5kW, 208V
(8) 1345VA hair dryers - 120V
Water heaters - 6kVA, 3-phase and 4kVA, 1-phase (208V)
(7) 1.2A exhaust fans - 120V
(10) 100W exit lights - 120V
1850VA Sign - 120V
(3) Heat Pumps - 9750VA, 3-phase
(1) Heat Pump - 2350VA, 1-phase (208V)
(3) Air Handlers (Heat) -12,000VA, 3-phase
(1) Air Handler (Heat) - 4800VA, 1-phase (208V)
(2) 120V Microwaves - 1800W
Icemaker - 1420VA (120V)
(2) Refrigerators - 1400VA (120V)
(3) Pedicure Units - 1440VA (120V)
(2) Manicure Units - 660VA (120V)
(3) Massage Units - 1056VA (120V)
(4) Shampoo Units - 680VA (120V
(9) Styling Units - 1877VA (120V)
120V Dishwasher - 1150VA
Receptacles - 76
(3) 120V Coffeemakers - 1kW
Sprinkler pump motor, 3-phase, 7.5HP
(10) 175W Metal Halide Fixtures 1.58A, 208V

The salon will also enclose a 234SF linen and supply closet. Determine the size main panelboard needed along with the service, neutral and grounding electrode conductors required to supply the salon. The general lighting load is less than 40 percent nonlinear.

1. GENERAL LIGHTING or ACTUAL LIGHTING LOADS

General Lighting Load <NEC References - 220.12 and Table 220.12>

Barber/Beauty Shop - 3965SF x 3VA/SF = 11,895VA
Closet - 234SF x ½VA/SF   =   117VA
                                        12,012VA

  A. General Lighting Load – 12,012VA x 1.25 = 15,015VA

Actual Lighting Load – NA

1.	LINE LOAD	NEUTRAL LOAD	
		Permitted Reduction	Prohibited Reduction
	15,015VA	12,012VA	--

2. OTHER LIGHTING LOADS

A. Sign/Outline (S/O) Lighting <NEC References - 220.12(F) and 600.5(A)>

Sign (120V) = 1850VA

Total (Sign/Outline Lighting) = 1850VA

B. Outside Lighting <NEC Reference - 220.18(B)>

Metal Halide - 208V x 1.58A x 10 = 3286.4VA

Total (Outside Lighting) = 3286.4VA

C. and D. NA

E. Miscellaneous (Write-ins. List individual voltage rating of each lighting load)

Exit lights (120V) - 100W x 10 = 1000W(VA)

Total (Miscellaneous) = 1000VA

F. Other Lighting (LINE and NEUTRAL) Loads Total [Add lines (A.) - (E.) where applicable]

Other Lighting Loads Total = 6136.4VA

TOTAL = 6136.4VA x 1.25 = 7670.5VA

2.	LINE LOAD	NEUTRAL LOAD	
		(120V)	
		Permitted Reduction	Prohibited Reduction
	7670.5VA	2850VA	--

3. RECEPTACLE LOADS

A(1) - NA
A(2) Non-continuous duty - General Purpose Receptacles <NEC 220.14(I)>

76 x 180VA = 13,680VA

APPLY DEMAND FACTORS

a. First 10,000VA (10kVA) (@100%) = 10,000VA
b. Remainder - 3680VA x .50   = <u>1840VA</u>

Total = 11,840VA

B. and C. NA

D. Total Receptacle (NEUTRAL and LINE) Load

Total Receptacle loads = 11,840VA

3. <u>LINE LOAD</u>   <u>NEUTRAL LOAD</u>
                                    Permitted   Prohibited
                                    Reduction   Reduction

11,840VA   11,840VA   --

4. KITCHEN EQUIPMENT - NA

5. SPECIFIC LOADS

Type Load	Calculation	
120V		
Coffeemakers - 1000W (VA) x 3	=	3000W (VA)
Dishwasher	=	1150VA
Hair Dryers - 1345VA x 8	=	10,760VA
Icemaker	=	1420VA
Manicure units - 660VA x 2	=	1320VA
Massage units - 1056VA x 3	=	3168VA
Microwaves - 1800VA x 2	=	3600VA
Pedicure units - 1440VA x 3	=	4320VA
Refrigerators - 1400VA x 2	=	2800VA
Shampoo units - 680VA x 4	=	2720VA
Styling units - 1877VA x 9	=	16,893VA
Washers - 1500VA x 2	=	3000VA
208V-1ϕ		
Dryers - 5500W (VA) x 2	=	11,000VA
Water Heater	=	4000W (VA)
208V-3ϕ		
Water Heater	=	<u>6000W (VA)</u>
	TOTAL =	75,151VA

5.	LINE LOAD	NEUTRAL LOAD	
		Permitted Reduction	Prohibited Reduction
	75,151VA	54,151VA	--

6. MOTOR LOADS

A. Continuous Duty

Motor Load	Calculation		
208V-3φ			
Sprinkler pump (7.5HP)	208V x 24.2A* x 1.732	=	8718.2VA
*Table 430.250			

B. Other Than Continuous Duty

Motor Load	Calculation		
120V			
Exhaust fans	120V x 1.2A x 7	=	1008.0VA

C. and D. NA

E. Total Motor Loads - [LINE LOAD - Add lines A. and B.] - [NEUTRAL LOAD - Total motor loads with neutral connections (120V)]

TOTAL = 9726.2VA

6.	LINE LOAD	NEUTRAL LOAD	
		Permitted Reduction	Prohibited Reduction
	9726.2VA	1008.0VA	--

7. MEDICAL EQUIPMENT - NA

8. INDUSTRIAL EQUIPMENT - NA

9. HEATING and AIR-CONDITIONING (AC) EQUIPMENT <NEC References - 220.50, 220.51, 220.60, 430.6(A)(1) and 440.6(A)>

HEAT PUMP (HP) UNITS

Heat Pumps Units	Calculation
(3φ) 9750VA Units (3)	9750VA x 3 = 29,250VA
(3φ) Air Handlers [Heat] (3)	12,000VA x 3 = 36,000VA

(1φ) 2350VA Units	= 2350VA
(1φ) Air Handlers [Heat]	= <u>4800VA</u>

TOTAL HEAT PUMP = 72,400VA

Line Load = 72,400VA

9.   <u>LINE LOAD</u>

<u>NEUTRAL LOAD</u>

Permitted Reduction	Prohibited Reduction

72,400VA   0   --

10. LARGEST MOTOR

Sprinkler pump motor (7.5HP - 24.2A/3φ - 8718.2VA) (highest current per NEC 430.17)
8718.2VA x .25 (25 percent) = 2179.6VA

10.   <u>LINE LOAD</u>

<u>NEUTRAL LOAD</u>

Permitted Reduction	Prohibited Reduction

2179.6VA   0   --

TOTAL DEMAND LOAD (LINE and NEUTRAL) (List each computed line and neutral loads below and total lines 1. – 10.)

	LINE LOAD	NEUTRAL LOAD Permitted Reduction	Prohibited Reduction
1. General Lighting	15,015.0VA	12,012.0VA	--
2. Other Lighting Loads	7,670.5VA	2,850.0VA	--
3. Receptacle Loads	11,840.0VA	11,840.0VA	--
4. Kitchen Equipment	--	--	--
5. Specific Loads	75,151.0VA	54,151.0VA	--
6. Motor Loads	9,726.2VA	1,008.0VA	--
7. Medical Equipment	--	--	--
8. Industrial Equipment	--	--	--
9. Heating and AC Equip.	72,400.0VA	0	--
10. Largest Motor	**2,179.6VA**	**0**	--

TOTAL = 193,982.3VA   81,861.0VA   --

OCCUPANCY'S OPERATING LINE VOLTAGE - <u>208</u>V(3φ)
(Given operating voltage or as determined per test examination)

CALCULATE MINIMUM LINE and NEUTRAL LOADS
(Divide Total Demand Load [VA] by operating line voltage [V])

LINE LOAD = 193,892.3VA / 208V x 1.732 = <u>538.46A</u>

NEUTRAL LOAD
   Permitted = 81,861VA / 208V x 1.732 = ~~227.23~~* 219.1A

*Where the neutral (permitted) load exceeds 200A, NEC 220.61(B)(2) permits the load to be reduced by 70 percent. Complete the following to determine the Total Neutral Load.

(1) 227.23A – 200A = 27.23A x .70 = 19.1A

(2) 19.1A + 200A = 219.1A

Total Neutral Load = <u>219.1A</u>

SIZE SERVICE (Size of service based on the calculated LINE LOAD)

SIZE SERVICE REQUIRED (minimum) <u>600A</u>
(Single rating or combination - Main overcurrent device(s) to total 600A)

The conductors or combinations selected for this installation is limited only to this question. Conductor ampacity derating is not applied in sizing the service and neutral conductors.

SIZING FEEDER/SERVICE CONDUCTORS

Applying NEC 310.10(H)(1) and Table 310.15(B)(16), as a minimum based on NEC 240.4(B), use two 350 kcmil copper conductors (75°C) per phase which are individually rated for 285A, as the service conductors based on the 600A service rating (600A/2 = 300A).

FEEDER/SERVICE CONDUCTORS <u>2-350 kcmil copper conductors per phase</u>

SIZING NEUTRAL CONDUCTOR(S)

Based on the calculated neutral load and the use of parallel service conductors, the neutral conductors must also be installed in parallel per NEC 310.10(H)(1). Because the service conductors will consist of two parallel conductors per phase, the neutral conductors must also consist of two parallel conductors based on the calculated neutral load being divided by two which results to 109.55A (219.1A / 2). Per Table 310.15(B)(16), at 75°C, a 2 AWG copper conductor which has a rated ampacity of 115A will satisfy the calculated neutral load being distributed amongst two such conductors. However, because NEC 310.10(H)(1) only allows a 1/0 AWG conductor or larger to be installed in parallel, two 1/0 AWG copper conductors must be used instead as the parallel neutral conductors.

NEUTRAL CONDUCTOR(S) <u>2-1/0 AWG copper conductors</u>

SIZING GROUNDING ELECTRODE CONDUCTOR

Based on the equivalent area of the selected parallel copper service conductors (700 kcmil-[350kcmil x 2]), per Table 250.66, a 2/0 AWG copper grounding electrode conductor must be used.

Although NEC 250.24(C)(1) requires the neutral (grounded) conductor to be either the same size or larger than the grounding electrode conductors, the total circular mils of the two 1/0 AWG neutral conductors (211,200 cmils [105,600 cmils x 2]) exceeds the circular mils of a 2/0 AWG conductor (133,100 cmils) per Table 8 of Chapter 9. As a result, this classifies the use of the two 1/0 AWG parallel neutral conductors as being larger than the 2/0 AWG grounding electrode conductor.

GROUNDING ELECTRODE CONDUCTOR 1-2/0 copper

Church [2]

93. A church building will consist of the following occupancies:

    5650SF sanctuary
    725SF nursery
    3100SF classrooms and office space
    950SF hallway
    575SF stairway
    1100SF storage space

    Determine the general lighting load based on the given occupancies. With the exception of the storage space assume the lighting loads in all occupancies to be continuous.

1. GENERAL LIGHTING or ACTUAL LIGHTING LOADS

General Lighting Load <NEC References - 220.12 and Table 220.12>

                5650SF sanctuary
                 725SF nursery
                3100SF classroom and office space
                9475SF

Because the sanctuary, nursery, classrooms and office space will be inclusive of the church the total square footage of these occupancies are totaled and applied with the unit load of a church. Therefore,

$$9475SF \times 1VA/SF = 9475VA$$

Since separate unit loads are listed in Table 220.12 for the remaining occupancies the general lighting loads will be determined accordingly. Therefore,

$$(\text{Hallway}) \quad 950\text{SF} \times \tfrac{1}{2}\text{VA/SF} = 475.0\text{VA}$$
$$(\text{Stairway}) \quad 575\text{SF} \times \tfrac{1}{2}\text{VA/SF} = 287.5\text{VA}$$
$$(\text{Storage space}) \quad 1100\text{SF} \times \tfrac{1}{4}\text{VA/SF} = \underline{275.0\text{VA}}$$
$$1037.5\text{VA}$$

Based on the requirements of this question and NEC 230.42(A), the general lighting load of the given occupancies is,

$$(9475\text{VA} + 475\text{VA} + 287.5) \times 1.25 + (275\text{VA}) = 13{,}071.9\text{VA}$$

Another way of determining other than the minimum general lighting load for this question is to consider the total of all occupancies and applying only the unit load of the church since such unit load meets and exceeds the minimum unit loads of the hallway, stairway and storage space occupancies.

94. A 240/120V, 1ϕ underground service will supply a church where the outside dimensions of the church are 130' x 130'.

The church's interior lighting will consist of the following fixtures:

120V
1 - Pendant Chandelier - 7.8A
120 - Lay in (4-lamp) trouffers w/electronic T8 ballast @ .68A
15 - 250W high bay fixtures - .83A
35 - 70W wall sconce fixtures
52 - 100W recessed cans

The remaining loads and other occupancies will consist of the following:

120V	**240V**
*Dishwasher - 8.3A	Heating units (4) - 20kW
100W Exit lights (17)	Heating units (3) - 15kW
*Garbage Disposal - 680VA	5-ton AC compressor (4) - 8250VA
Ceiling Fans (8) - 1.34A	4-ton AC compressor (3) - 7570VA
*Freezer - 13.8A	¾HP Air Handlers (7)
175W Metal Halide (11) - .94A (outside)	*Water Heater - 4.5kW
*Microwave Oven - 2000W	*Water Heater - 6.5kW
Receptacles - 203 (non-continuous)	
*Refrigerator - 18.4A	**240/120V**
Sign - 11.3A	
Soda Machine - 1480VA	*Dryer - Heat 23.7A/ Motor 5A
Vent Fans (3) - 2.2A	*Cooking Range - 15kW
*Washer -11.5A	(Neutral 10 percent of Range load)
Water Cooler (2) - 4.2A	

*Commercial rated appliances/equipment

What size service is required to serve the church? Using THWN copper conductors, what size service, neutral and grounding electrode conductors are needed to supply the service? Consider the use of 3 parallel conductors per line and neutral.

1. GENERAL LIGHTING or ACTUAL LIGHTING LOADS

General Lighting Load <NEC References - 220.12 and Table 220.12>

Church - 130ft. x 130ft. x 1VA/SF = 16,900VA
Closets - 200SF x ½VA/SF   =   100VA
Stairways – 144SF x ½VA/SF   =   72VA
                                    17,072VA

   A. General Lighting Load - ~~17,072VA~~ <Omit General Lighting Load - Actual Lighting Load larger>

Actual Lighting Load

Type Fixture	VA rating		No. of Fixtures		TOTAL VA
Church					
Chandelier	120V x 7.8A	x	1	=	936
Trouffers	120V x .68A	x	120	=	9,792
250W high bay	120V x .83A	x	15	=	1,494
Wall sconces	70W	x	35	=	2,450
Recessed Cans	100W	x	52	=	5,200
					19,872

   B. Actual Lighting Load - 19,872VA x 1.25 = 24,840VA

1.	LINE LOAD		NEUTRAL LOAD	
			Permitted Reduction	Prohibited Reduction
	24,840VA		19,872VA	--

2. OTHER LIGHTING LOADS

   A. Sign/Outline (S/O) Lighting <NEC References - 220.12(F) and 600.5(A)>
       Sign - 120V x 11.3A = 1356VA
     Total (Sign/Outline Lighting) = 1356VA

B. Outside Lighting <NEC Reference - 220.18(B)>

Metal Halide - 120V x .94A x 11 = 1240.8VA

Total (Outside Lighting) = 1240.8VA

C. and D. NA

E. Miscellaneous

(120V) Exit lights - 100W x 17 = 1700VA

Total (Miscellaneous) = 1700VA

F. Other Lighting (LINE and NEUTRAL) Loads Total [Add lines A. - E. where applicable]

Other Lighting Loads Total = 4296.8VA

TOTAL = 4296.8VA x 1.25 = 5371VA

2.	LINE LOAD	NEUTRAL LOAD

		(120V)	
		Permitted Reduction	Prohibited Reduction
	5371VA	4296.8VA	--

3. RECEPTACLE LOADS

A(1) - NA

A(2) Non-continuous duty - General Purpose Receptacles <NEC 220.14(I)>

203 x 180VA = 36,540VA

APPLY DEMAND FACTORS

(1) First 10,000VA (10kVA) (@100%) = 10,000VA

(2) Remainder - 26,540VA x .50   = 13,270VA

Total = 23,270VA

B. NA

C. NA

D. Total Receptacle (NEUTRAL and LINE) Load

Total Receptacle loads = 23,270VA

3.   <u>LINE LOAD</u>

	<u>NEUTRAL LOAD</u>	
	Permitted Reduction	Prohibited Reduction

      23,270VA   23,270VA   --

4. KITCHEN EQUIPMENT

Type Kitchen   Calculation

Type Kitchen	Calculation	
Cooking Range	=	15,000W (VA)
Dishwasher	120V x 8.3A =	996VA
Freezer	120V x 13.8A =	1656VA
Garbage Disposal	=	680VA
Microwave	=	2000VA
Refrigerator	120V x 18.4A =	<u>2208VA</u>
	Total (6) =	22,540VA

a. Per Table 220.56 - 22,540VA x .65 = 14,651VA
b. Two Largest Kitchen Loads -15,000VA (Range) + 2208VA (Refrigerator) = 17,208VA

KITCHEN LOAD (Larger of a. and b.) = 17,208VA

NEUTRAL LOAD (120V and 240/120V) (6)
7540VA (5) + (15,000VA x .10) x .65 = 5876VA

4.   <u>LINE LOAD</u>

	<u>NEUTRAL LOAD</u>	
	Permitted Reduction	Prohibited Reduction

      17,208VA   5876VA   --

5. SPECIFIC LOADS

Type Load	Calculation	
240V		
Dryer (Heat)	240V x 23.7A =	5688VA
Water Heater	4500W (VA) =	4500VA
Water Heater	6500W (VA) =	<u>6500VA</u>
	TOTAL =	16,688VA

5. <u>LINE LOAD</u>   <u>NEUTRAL LOAD</u>

                                              Permitted   Prohibited
                                              Reduction   Reduction

    16,688VA   0   --

6. MOTOR LOADS

A. NA

B. Other Than Continuous Duty

Motor Load	Calculation		
120V			
Ceiling Fans (8)	120V x 1.34 x 8	=	1286.4VA
Soda Machine		=	1480.0VA
Vent Fans (3)	120V x 2.2A x 3	=	792.0VA
Washer	120V x 11.5A	=	1380.0VA
Water Coolers (2)	120V x 4.2A x 2	=	1008.0VA
			5946.4VA
240V			
Dryer	240V x 5A	=	1200.0VA

C. NA

D. NA

E. Total Motor Loads - [LINE LOAD - Add line B.] - [NEUTRAL LOAD - Total motor loads with neutral connections (120V)]

                                              TOTAL =   7146.4VA

6. <u>LINE LOAD</u>   <u>NEUTRAL LOAD</u>

                                              Permitted   Prohibited
                                              Reduction   Reduction

    7146.4VA   5946.4VA   --

7. MEDICAL EQUIPMENT - NA

8. INDUSTRIAL EQUIPMENT - NA

9. HEATING and **AIR-CONDITIONING (AC) EQUIPMENT** <NEC References - 220.50, 220.51, 220.60, 430.6(A)(1) and 440.6(A)>

ELECTRICAL HEATING UNITS

Heating Unit	Calculation
20kW Units (4)	20,000W x 4 = 80,000(W) VA
15kW Units (3)	15,000W x 3 = 45,000(W) VA

TOTAL HEAT = 125,000VA

AIR-CONDITIONING (AC) UNITS

AC Unit	Calculation
5-ton Units (4)	8250VA x 4 = 33,000VA
4-ton Units (3)	7570VA x 3 = 22,710VA

TOTAL AC = 55,710VA

Line Load = 125,000VA + (11,592VA) = 136,592VA
        (Largest Load)   (Air Handlers)*

¾HP - 240V x 6.9A* x 7 = 11,592VA
*Tables 430.248

9.	LINE LOAD	NEUTRAL LOAD	
		Permitted Reduction	Prohibited Reduction
	136,592VA	0	--

10. LARGEST MOTOR

Refrigerator (motor with highest current per NEC 430.17 and 440.7)
2208VA x .25 (25 percent) = 552VA

10.	LINE LOAD	NEUTRAL LOAD	
		Permitted Reduction	Prohibited Reduction
	552VA	552VA	--

TOTAL DEMAND LOAD (LINE and NEUTRAL) (List each computed line and neutral loads below and total lines 1. – 10.)

	LINE LOAD	NEUTRAL LOAD	
		Permitted Reduction	Prohibited Reduction
1. General Lighting	24,840.0VA	19,872.0VA	--
2. Other Lighting Loads	5371.0VA	4296.8VA	--
3. Receptacle Loads	23,270.0VA	23,270.0VA	--
4. Kitchen Equipment	17,208.0VA	5876.0VA	--
5. Specific Loads	16,688.0VA	0	--
6. Motor Loads	7146.4VA	5946.4VA	--
7. Medical Equipment	--	--	--
8. Industrial Equipment	--	--	--
9. Heating and AC Equip.	136,592.0VA	0	--
10. Largest Motor	**552.0VA**	**552.0VA**	--
TOTAL =	231,667.4VA	59,813.2VA	--

OCCUPANCY'S OPERATING LINE VOLTAGE - 240V
(Given operating voltage or as determined per test examination)

CALCULATE MINIMUM LINE and NEUTRAL LOADS
(Divide Total Demand Load [**VA**] by operating line voltage [**V**])

LINE LOAD = 231,667.4VA / 240V = 965.28A

NEUTRAL LOAD - Permitted = 59,813.2VA / 240V = ~~249.22A~~* 234.45A

*Where the neutral (permitted) load exceeds 200A, NEC 220.61(B)(2) permits the load to be reduced by 70 percent. Complete the following to determine the Total Neutral Load.

(1) 249.22A – 200A = 49.22A x .70 = 34.45A

(2) 34.45A + 200A = 234.45A

Total Neutral Load = 234.45A

SIZE SERVICE (Size of service based on the calculated LINE LOAD)

SIZE SERVICE REQUIRED (minimum) 1000A
(Single rating or combination - Main overcurrent device(s) to total 1000A)

The conductors or combinations selected for this installation is limited only to this question. Conductor ampacity derating is not applied in sizing the service and neutral conductors.

SIZING FEEDER/SERVICE CONDUCTORS

Applying NEC 310.10(H)(1) and Table 310.15(B)(16), as a minimum based on NEC 240.4(C), use three 400 kcmil copper conductors (75°C) per phase which are individually rated for 335A, as the service conductors based on the 1000A service rating (1000A/3 = 333.33A).

FEEDER/SERVICE CONDUCTORS <u>3-400 kcmil copper conductors per phase</u>

SIZING NEUTRAL CONDUCTOR(S)

Based on the calculated neutral load and the use of parallel service conductors, the neutral conductors must also be installed in parallel per NEC 310.10(H)(1). Because the service conductors will consist of three parallel conductors per phase, the neutral conductors must also consist of three parallel conductors based on the calculated neutral load being divided by three which results to 78.15A (234.45A / 3). Per Table 310.15(B)(16), at 75°C, a 4 AWG copper conductor which has a rated ampacity of 85A will satisfy the calculated neutral load being distributed amongst three such conductors. However, because NEC 310.10(H)(1) only allows a 1/0 AWG conductor or larger to be installed in parallel, three 1/0 AWG copper conductors must be used instead as the parallel neutral conductors.

NEUTRAL CONDUCTOR(S) <u>3-1/0 AWG copper conductors</u>

SIZING GROUNDING ELECTRODE CONDUCTOR

Based on the equivalent area of the selected parallel copper service conductors (1200 kcmil), per Table 250.66, a 3/0 AWG copper grounding electrode conductor must be used. Where the equivalent area for parallel copper service-entrance conductors exceeds 1100 kcmil, the grounding electrode conductor is not required to be larger than 3/0 AWG copper for any installation.

GROUNDING ELECTRODE CONDUCTOR <u>1-3/0 AWG copper conductor</u>

Hospital

Before performing the load calculation for the hospital below consider the following. Based on the definitions of NEC 517.2 an *essential electrical system* for a hospital in abbreviated terms must have the capabilities to provide an adequate supply of electrical power deemed essential to maintain life, safety and usual operations during the disruption or loss of normal electrical power regardless of reason.

Referring to the illustrations provided in Article 517 an essential electrical system in a hospital consists of an emergency system and an equipment system. According to NEC 517.30(B)(2) the *emergency system* must be limited to circuits essential to life safety and critical patient care which are designated as the *life safety branch* and the *critical branch*. According to NEC 517.30(B)(3) the *equipment system* must supply major electrical equipment necessary for patient care and basic hospital operations.

To best understand the critical and life safety branches of an emergency system and the equipment system which identifies the essential electrical system of a hospital the definitions of such terms are provided in NEC 517.2.

Critical Branch - A subsystem of the *emergency system* consisting of feeders and branch circuits supplying energy to task illumination, special power circuits, and selected receptacles serving areas and functions related to patient care and that are connected to alternate power sources by one or more transfer switches during interruption of normal power source.

Life Safety Branch - A subsystem of the *emergency system* consisting of feeders and branch circuits, meeting the requirements of Article 700 (Emergency Systems) and intended to provide adequate power needs to ensure safety to patients and personnel, and that are automatically connected to alternate power sources during interruption of the normal power source.

Equipment System - A system of circuits and equipment arranged for delayed, automatic, or manual connection to the alternate power source and that serves primarily 3-phase power equipment.

95. A 385,000SF hospital is supplied from a 2.4kV three-phase source. The voltage is stepped down to be utilized at 480/277V-3φ. Although there are other voltage applications in this hospital that are less than 480/277V-3φ they are only vaguely mentioned. If switchgear was used to receive the hospital's incoming voltage what size equipment is required?

Emergency System
Critical Branch - 310kVA
Life Safety Branch - 295kVA

Equipment System
Motors 480V - 3φ
  Dumbwaiters - 75,286VA
  Freight, Patients, Passengers Elevators (34) - (968,437VA)
  Escalators - (413,423VA)
  Water, Fire and Pressure Pumps - (853,637VA)
  Largest Motor - 150,653VA
  Medical Equipment
    142 - X-ray units (P-Primary, S-Secondary)
    (momentary rating) - 1φ/P-208V/.165A@S-90kV
    (long-time rating)   - 1φ/P-208V/.022A@S-225kV
Heating and Air-Conditioning 480V - 3φ
  Heating - 1.3MW (Hospital heat -75 percent electric/25 percent gas)
  AC - 1.1MVA
Communications/Data Systems (nonlinear) (18 percent of load at 120V) - 37,678VA
Outside Lighting 277V - 81,023VA

Other Loads
Hospital Operations
  General purpose receptacles (120V) - 6349
Maintenance shop - 23,280VA
Cafeteria cooking equipment - 60 commercial appliances @ 573,851VA (20 percent 120V appliances)
  All cooking equipment thermostatically controlled
  Two Largest cooking equipment loads - 35,758VA (over 208V)
Signs 277V - 10,000VA

Other Occupancies
Corridors - 32,000SF
Stairways - 25,000SF

1. GENERAL LIGHTING or ACTUAL LIGHTING LOADS

General Lighting Load <NEC References – 220.12 and Table 220.12>

Corridors - 32,000SF x ½VA/SF	=	16,000VA*
Stairways - 25,000SF x ½VA/SF	=	12,500VA*
Hospital - 385,000SF x 2VA/SF	=	770,000VA

*loads expected to operate continuous

APPLY DEMAND FACTORS FOR HOSPITALS ONLY

a. First 50,000VA or less of above TOTAL (at 40%)

    50,000VA x .40   =   20,000VA

b. 720,000VA x .20 (at 20%)   = 144,000VA
    (Remainder of TOTAL VA exceeding 50,000VA)**

    **(770,000VA [TOTAL] – 50,000VA = 720,000VA)

                    TOTAL (Lines a and b)   = 164,000VA
                                            (Derated Lighting Load)

A. General Lighting Load - (16,000VA + 12,500VA) x 1.25 + 164,000VA = 199,625VA

    Actual Lighting Load - NA

1.	LINE LOAD	NEUTRAL LOAD	
		Permitted Reduction	Prohibited Reduction
	199,625VA	--	192,500VA

2. OTHER LIGHTING LOADS

A. Sign/Outline (S/O) Lighting <NEC References - 220.12(F) and 600.5(A)>

Signs (277V) = 10,000VA

Total (Sign/Outline Lighting) = 10,000VA

B. Outside Lighting <NEC Reference - 220.18(B)>

Outside parking lights (277V) - 81,023VA

Total (Outside Lighting) = 81,023VA

C. - E. NA

F. Other Lighting (LINE and NEUTRAL) Loads Total [Add lines A. and B. where applicable]

Other Lighting Loads Total = 91,023VA

TOTAL = 91,023VA x 1.25 = 113,778.75VA

2.	LINE LOAD	NEUTRAL LOAD	
		Permitted Reduction	Prohibited Reduction
	113,778.75VA	--	91,023VA

3. RECEPTACLE LOADS

A(1) - NA

A(2) Non-continuous duty - General Purpose Receptacles <NEC 220.14(I)>

6349 x 180VA = 1,142,820VA

APPLY DEMAND FACTORS
a. First 10000VA (10kVA) (@100%)	=	10,000VA
b. Remainder - 1,132,820VA x .50	=	566,410VA
Total =		576,410VA

B. and C. - NA

D. Total Receptacle (NEUTRAL and LINE) Load

Total Receptacle loads = 576,410VA

3. **LINE LOAD**

    576,410VA

	NEUTRAL LOAD	
	Permitted Reduction	Prohibited Reduction
	576,410VA	--

4. KITCHEN EQUIPMENT

a. Per Table 220.56 - (48) 573,851VA x .80 x .65 = 298,402.52VA
b. Two Largest Kitchen Loads - 35,758VA

KITCHEN LOAD (Larger of a. and b.) = 298,402.52VA

NEUTRAL LOAD (120V) - (12) 573,851VA x .20 x .65 = 74,600.63VA

4. **LINE LOAD**

    298,402.52VA

	NEUTRAL LOAD	
	Permitted Reduction	Prohibited Reduction
	74,600.63VA	--

5. SPECIFIC LOADS

Type Load		Calculation
Communications/Data Systems (37,678VA)	=	30,895.96VA*
Critical Branch	=	310,000.00VA
Life Safety Branch	=	295,000.00VA
Maintenance shop	=	23,280.00VA
TOTAL =		659,175.96VA

*Line Load = 30,895.96VA (37,678VA x .82) Neutral Load = 6782.04VA (37,678VA x .18)

5. **LINE LOAD**

    659,175.96VA

	NEUTRAL LOAD	
	Permitted Reduction	Prohibited Reduction
	--	6782.04VA

6. MOTOR LOADS

A. Continuous Duty

Motor Load	Calculation		
Escalators	--	=	413,423.00VA
Water, Fire and Pressure Pumps	--	=	853,637.00VA

B. and C. NA

D. Intermittent Duty Cycle

Motor Load	Calculation		
Dumbwaiters Freight, Patients, Passengers Elevators (34) - 968,437VA	--	=	75,286.00VA
968,437VA x .72*		=	697,274.64VA

*Feeder Demand Factor for 34 Elevators @ .72 - Table 620.14 (assumed - not under constant load)

E. Total Motor Loads - [LINE LOAD - Add lines A. and D.] - [NEUTRAL LOAD - Total motor loads with neutral connections (120V)]

$$TOTAL = 2,039,620.64VA$$

6. **LINE LOAD**

NEUTRAL LOAD
Permitted   Prohibited
Reduction   Reduction

2,039,620.64VA   --   --

7. MEDICAL EQUIPMENT

A. Diagnostic Equipment

(1) Branch-Circuits Load <NEC Reference - 517.73(A)(1)> (Individual branch-circuit load based on either the momentary or long-time rating of the given load.) (Use the greater of the two ratings per circuit).

142 X-ray units
(momentary rating) - 1ϕ/P-208V/.165A@S-90kV
(long-time rating)   - 1ϕ/P-208V/.022A@S-225kV

Momentary-time Primary Current (PC)/VA

PC = (90,000V/208V) x .165A = 71.4A
VA = 208V x 71.4A = 14,851.2VA

Long-time Primary Current(PC)/VA

PC = (225,000V/208V) x .022A = 23.8A
VA = 208V x 23.8A = 4950.4VA

(2) Feeder Load <NEC Reference - 517.73(A)(2)> (Two or more branch circuits supplied from a common or separate feeders.)

Feeder	Calculation		
Largest Unit	14,851.2VA x .50	=	7425.6VA
Next Largest Unit	14,851.2VA x .25	=	3712.8VA
Additional Units	14,851.2VA x 140 x .10	=	207,916.8VA
		TOTAL =	219,055.2VA

7. **LINE LOAD**

	NEUTRAL LOAD	
	Permitted Reduction	Prohibited Reduction
219,055.2VA	--	--

8. INDUSTRIAL EQUIPMENT - NA

9. HEATING and AIR-CONDITIONING (AC) EQUIPMENT <NEC References - 220.50, 220.51, 220.60, 430.6(A)(1) and 440.6(A)>

Heating - 1.3MW (1,300,000(W) VA) (Larger)
AC   - 1.1MVA (1,100,000VA)

9. **LINE LOAD**

	NEUTRAL LOAD	
	Permitted Reduction	Prohibited Reduction
1,300,000VA	0	--

10. LARGEST MOTOR

150,653VA x .25 (25percent) = 37,663.25VA

10. **LINE LOAD**

	NEUTRAL LOAD	
	Permitted Reduction	Prohibited Reduction
37,663.25VA	--	--

TOTAL DEMAND LOAD (LINE and NEUTRAL) (List each computed line and neutral loads below and total lines 1. – 10.) Highlights identify loads affected by re-calculations.

	LINE LOAD	NEUTRAL LOAD	
		Permitted Reduction	Prohibited Reduction
1. General Lighting	199,625.00VA	--	192,500.00VA
2. Other Lighting Loads	113,778.75VA	--	91,023.00VA
3. Receptacle Loads	576,410.00VA	576,410.00VA	--
4. Kitchen Equipment	298,402.52VA	74,600.63VA	--
5. Specific Loads	659,175.96VA	--	6782.04VA
6. Motor Loads	2,039,620.64VA	--	--
7. Medical Equipment	219,055.20VA	--	--
8. Industrial Equipment	--	--	--
9. Heating and AC Equip.	1,300,000.00VA	--	--
10. Largest Motor	37,663.25VA	--	--
TOTAL =	5,443,713.32VA	651,010.63VA	290,305.04VA

OCCUPANCY'S OPERATING LINE VOLTAGE - 2400V
(Given operating voltage or as determined per test examination)

CALCULATE MINIMUM LINE LOAD
(Divide Total Demand Load **[VA]** by operating line voltage **[V]**)

LINE LOAD = 5,443,713.32VA / 2400V x 1.732 = 1309.59A

SIZE SWITCHGEAR (Size of switchgear based on the calculated LINE LOAD)

SIZE SWITCHGEAR REQUIRED (minimum) 1600A

REMINDER - Although various loads of this question were without an assigned voltage or phase reference, the overall load calculation could still be performed. The only distinction required was that of a line load opposed to a neutral load. Again, the most important thing to remember in a situation like this is to always use the highest voltage given based upon the need to size a desired service or service equipment. All other load voltages will be included which will at some point require separate calculations for sizing transformers to accommodate such voltage and phase applications.

Hotel/Motel [2]

For "Hotel/Motel" reference format see **(Worksheet H - Volume 4)** STANDARD LOAD CALCULATION FOR HOTELS and MOTELS.

96. An eighty unit motel accommodation is served by a 208/120V, 3-phase, 4-wire service. The lay-out of the complex consists of 40-285SF units and 40-325SF units. Each unit and the motel's overall electrical system will consist of the following loads:

Unit Rooms
5kW heat pumps with supplemental heat- 208V, 1-phase
1500W hair dryer (120V)
1000W tanning light (120V)
7 - duplex receptacles

Motel Office
550SF (continuous use)
6.2kW heat pump with supplemental heat- 208V, 1-phase
1.3kW heat-vent-light (120V)
12 - duplex receptacles
8.7A copy machine- 208V, 1-phase

Laundry Room
8 - 4 lamp fluorescent fixtures (nonlinear) - 1.14A (120V)
5kW heat pump with supplemental heat- 208V, 1-phase
2 - 7.5kW electric dryers - 208V, 1-phase
3 - ¾ HP washers (208V, 1-phase)
1 - 3.8kW steam presser (208V, 1-phase)
7 - duplex receptacles

Other Loads
4 - 1375VA soda machines (120V)
4 - 985VA snack machines (120V)
6 - 1255VA ice machines (120V)
2 - Signs, 8.67A (nonlinear) (120V)
10kVA outside lighting (nonlinear) - 208V, 1-phase (continuous)
2 - 25HP water pumps, 3-phase (208V)
3 - 16kW electric boilers (208V)
40HP fire pump, 3-phase (208V)
3HP pressure pump, 3-phase (208V)

Considering the motel's general lighting load is linear determine the motel's service and neutral loads.

1. GENERAL LIGHTING and RECEPTACLE LOADS <NEC References - 220.12, Table 220.12, 220.14(J), 220.42 and Table 220.42> Where actual lighting loads are applied receptaccle loads should be calculated separately. Where receptacle loads are calculated separately or considered continuous refer to itme **3**. Receptacles Load.

General Lighting and Receptacle Loads (Room receptacles included)

$$285SF \times 40 \times 2VA/SF = 22,800VA$$
$$325SF \times 40 \times 2VA/SF = \underline{26,000VA}$$
A. General Lighting and Receptacle Loads $= 48,800VA$

Actual Lighting Load - NA

APPLY DEMAND FACTORS <NEC References - 220.42 and Table 220.42>

a. First 20,000VA or less of above TOTAL (at 50%)

20,000VA x .50 $= 10,000VA$

b. 28,800VA x .40 (at 40%) $= 11,520VA$
   (48,800VA – 20,00VA)

c. $= \underline{\quad 0 \quad}$

TOTAL (Lines a. - c.) $= 21,520VA$

GENERAL LIGHTING and RECEPTACLE LOADS

1. - 3. <u>LINE LOAD</u>	<u>NEUTRAL LOAD</u>	
	Permitted Reduction	Prohibited Reduction
21,520VA	21,520VA	--

2. OTHER LIGHTING LOADS

A. Sign/Outline (S/O) Lighting <NEC References - 220.12(F) and 600.5(A)>

Signs - 120V x 8.67A x 2 = 2080.8VA

Total (Sign/Outline Lighting) = 2080.8VA

B. Outside Lighting <NEC Reference - 220.18(B)>

Total (Outside Lighting) = 10,000VA (208V-1φ)

C. and D. NA

E. Miscellaneous

Fluorescent lights (Laundry Room) - 120V x 1.14A x 8 = 1094.4VA

F. Other Lighting (LINE and NEUTRAL) Loads Total [Add lines A., B. and E. where applicable]

Other Lighting Loads Total = 13,175.2VA

TOTAL = 13,175.2VA x 1.25 = 16,469VA

2.   <u>LINE LOAD</u>   <u>NEUTRAL LOAD</u>

	LINE LOAD	NEUTRAL LOAD	
		Permitted Reduction	Prohibited Reduction
	16,469VA	--	3175.2VA

3. RECEPTACLE LOADS

A(1) - NA

A(2) General Purpose Receptacles (Laundry Room)

7 x 180VA = 1260VA

B. Non-continuous duty (Motel Office) <NEC 220.14(K)>

(1) 12 x 180VA = 2160VA (Larger)

(2) 550SF x 1VA/SF = 550VA

C. - NA

D. Total Receptacle (NEUTRAL and LINE) Load
LINE and NEUTRAL = 1260VA + 2160VA = 3420V

3.   <u>LINE LOAD</u>   <u>NEUTRAL LOAD</u>

	LINE LOAD	NEUTRAL LOAD	
		Permitted Reduction	Prohibited Reduction
	3420VA	3420VA	--

4. OTHER LOADS (Continuous and Noncontinuous)

List continuous and noncontinuous loads and calculate volt-ampere (VA) ratings if not given.

Continuous Loads	Calculations		
Office Gen. Lighting (120V)	550SF x 3.5VA	=	1925VA
	TOTAL	=	1925VA

Noncontinuous Loads	Calculations		
Copy machine (208V)	208V x 8.7A	=	1809.6VA
Electric boilers	16,000VA x 3	=	48,000.0VA
Electric dryers (208V)	7500VA x 2	=	15,000.0VA
Fire pump (40HP)	208V x 114A** x 1.732	=	41,069.2VA
Hair dryers (120V)	1500VA x 80	=	120,000.0VA
Heat-vent-light (120V)		=	1300.0VA
Ice machines (120V)	1255VA x 6	=	7530.0VA
Pressure pump (3HP)	208V x 10.6A** x 1.732	=	3818.7VA
Soda machines (120V)	1375VA x 4	=	5500.0VA
Snack machines (120V)	985VA x 4	=	3940.0VA
Steam presser		=	3800.0VA
Tanning lights (120V)	1000VA x 80VA	=	80,000.0VA
Washers (¾ HP)	208V x 7.6A* x 3	=	4742.4VA
Water pumps (25HP)	208V x 74.8A** x 1.732 x 2 =		53,894.3VA

*Table 430.248   **Table 430.250

	TOTAL	=	390,404.2VA

Total Other Loads - 1925VA x 1.25 + 390,404.2 VA = 392,810.5VA

4.	LINE LOAD	NEUTRAL LOAD	
		(120V)	
		Permitted Reduction	Prohibited Reduction
	392,810.5VA	220,195VA	--

5. HEATING and **AIR-CONDITIONING (AC) EQUIPMENT** <NEC References - 220.50, 220.51, 220.60, 430.6(A)(1) and 440.6(A)>

HEAT PUMP (HP) UNITS

Heat Pump Unit	Calculation		
6.2kW Units	6200W	=	6200(W) VA
5kW Unit (81)	5000W x 81	=	405,000(W) VA
	TOTAL HEAT PUMPS	=	411,200VA

5.   <u>LINE LOAD</u>

<table>
<tr><td></td><td colspan="2"><u>NEUTRAL LOAD</u></td></tr>
<tr><td></td><td>Permitted
Reduction</td><td>Prohibited
Reduction</td></tr>
<tr><td>411,200VA</td><td>--</td><td>--</td></tr>
</table>

6. LARGEST MOTOR

Fire pump (motor with highest current per NEC 430.17 and 440.7)
41,069.2VA x .25(25percent) = 10,267.3VA

6.   <u>LINE LOAD</u>

<table>
<tr><td></td><td colspan="2"><u>NEUTRAL LOAD</u></td></tr>
<tr><td></td><td>Permitted
Reduction</td><td>Prohibited
Reduction</td></tr>
<tr><td>10,267.3VA</td><td>--</td><td>--</td></tr>
</table>

TOTAL DEMAND LOAD (LINE and NEUTRAL) (List each computed line and neutral loads below and total lines 1. - 6.)

	<u>LINE LOAD</u>	<u>NEUTRAL LOAD</u> Permitted Reduction	Prohibited Reduction
1. General Lighting	21,520.0VA	21,520VA	--
2. Other Lighting Loads	16,469.0VA	--	3175.2VA
3. Receptacle Loads	3420.0VA	3420VA	--
4. Other Loads	392,810.5VA	220,195VA	--
5. Heating and AC Equip.	411,200.0VA	--	--
6. Largest Motor	10,267.3VA	--	--
Total Demand Load (VA) =	855,686.8VA	245,135VA	3175.2VA

OCCUPANCY'S OPERATING LINE VOLTAGE - <u>208</u>V(3φ)
(Given operating voltage or as determined per test examination)

CALCULATE MINIMUM LINE and NEUTRAL LOADS
(Divide Total Demand Load [**VA**] by operating line voltage [**V**])

<u>LINE (SERVICE) LOAD</u> = 855,686.8VA / 208V x 1.732 = <u>2375.22A</u>

<u>NEUTRAL LOAD</u>

      Permitted = 245,135VA / 208V x 1.732 = ~~680.45A~~* <u>545.13A</u>

      Prohibited = 3175.2VA / 208V x 1.732 = <u>8.81A</u>

*Where the neutral (permitted) load exceeds 200A, NEC 220.61(B)(2) permits the load to be reduced by 70 percent. Complete the following to determine the Total Neutral Load.

(1) 680.45A − 200A = 480.45A x .70 = 336.32A

(2) 336.32A + 200A = 536.32A

Total Neutral Load = 536.32A + 8.81A = 545.13A

97. A five-story luxury hotel consisting of 400 guest rooms will be supplied by a 480/277V, 4 wire, 3-phase service. A need for four 3-phase, 4-wire, 208/120V feeders will be required to serve the guest rooms, house loads and other occupancies. Calculate the service load to determine the size service required for the hotel.

Guest Rooms

125 - 335SF
100 - 380SF
 90 - 400SF
 50 - 425SF
 35 - 470SF

House loads (continuous and noncontinuous loads are combined)

3φ-480/277V

(Elevators, swimming pool motors, boilers, parking and outside lighting, water pumps, fire pump and jockey motor, AC and heating) - 1,054,786VA (10 percent neutral loads - 5 percent nonlinear)
Largest Motor - 103,088.64VA

3φ-208/120V

(Interior and exterior lighting, receptacles, computers, telecommunication and office equipment, laundry needs, restaurant, soda and vending machines, signs, AC and heating) - 443,600VA (30 percent neutral loads - 10 percent nonlinear)

Other Occupancies (Linear lighting-continuous)

Hallways - 2685SF
Stairways - 1650SF

The guest room's general lighting load is 120V nonlinear.

For "Motel" reference format see **(Worksheet H** - Volume 4**)** STANDARD LOAD CALCULATION FOR HOTELS and MOTELS.

1. GENERAL LIGHTING and RECEPTACLE LOADS <NEC References - 220.12, Table 220.12, 220.14(J), 220.42 and Table 220.42> Where actual lighting loads are applied receptacle loads should be calculated separately. Where receptacle loads are calculated separately or considered continuous refer to Item **3**. Receptacles Load.

General Lighting and Receptacle Loads (Room receptacles included)

$$335SF \ x \ 125 \ x \ 2VA/SF \ = 83,750VA$$
$$380SF \ x \ 100 \ x \ 2VA/SF \ = 76,000VA$$
$$400SF \ x \ \ 90 \ x \ 2VA/SF \ = 72,000VA$$
$$425SF \ x \ \ 50 \ x \ 2VA/SF \ = 42,500VA$$
$$470SF \ x \ \ 35 \ x \ 2VA/SF \ = \underline{32,900VA}$$

    A. General Lighting and Receptacle Loads = 307,150VA

Actual Lighting Load – NA

APPLY DEMAND FACTORS <NEC References - 220.42 and Table 220.42>

a. First 20,000VA or less of above TOTAL (at 50%)

   20,000VA x .50   = 10,000VA

b. 80,000VA x .40 (at 40%)   = 32,000VA
   (Total VA – 20,001VA up to 100,000VA)

c. 207,150VA x .30 (at 30%)   = 62,145VA
   (Remainder of TOTAL VA exceeding 100,000VA)*

   *(307,150VA (TOTAL) – 100,000 VA = 207,150VA)

        TOTAL (Lines a. - c.)   = 104,145VA

GENERAL LIGHTING and RECEPTACLE LOADS

1. - 3. LINE LOAD	NEUTRAL LOAD (120V)	
	Permitted Reduction	Prohibited Reduction
104,145VA	--	104,145VA

2. OTHER LIGHTING LOADS - NA

3. RECEPTACLE LOADS - NA

4. OTHER LOADS (Continuous and Noncontinuous)

List continuous and noncontinuous loads and calculate volt-ampere (VA) ratings if not given.

Continuous Loads	Calculations		
Hallways	2685SF x .5VA/SF	=	1342.50VA*
	1342.50VA x 1.25VA	=	1678.13VA
Stairways	1650SF x .5VA/SF	=	825.00VA*
	825 x 1.25VA	=	1031.25VA

* Neutral Loads

House loads (3φ-480/277V)	Calculations		
Service	--		= 1,054,786.00VA
Neutral	1,054,786VA x .10	=	105,478.60VA
Nonlinear (Prohibited Neutral)	105,478.6VA x .05	=	5273.93VA
Permitted Neutral	105,478.6VA – 5273.93VA		100,204.67VA

House loads (3φ-208/120V)	Calculations		
Service	--		= 443,600.00VA
Neutral	443,600.00VA x .30	=	133,080.00VA
Nonlinear (Prohibited Neutral)	133,080VA x .10	=	13,308.00VA
Permitted Neutral	133,080VA – 13,308VA	=	119,772.00VA

Total House loads

Service	1678.13VA + 1031.25VA + 1,054,786.00VA + 443,600.00VA	= 1,501,095.38VA
Permitted Neutral	1342.5VA + 825VA + 100,204.67VA + 119,772.00VA	= 222,144.17VA
Prohibited Neutral	5273.93VA + 13,308.00VA	= 18,581.93VA

4.	LINE LOAD	NEUTRAL LOAD

(120V and 277V)

	Permitted Reduction	Prohibited Reduction
1,501,095.38VA	222,144.17VA	18,581.93VA

5. HEATING and AIR-CONDITIONING (AC) EQUIPMENT - NA

6. LARGEST MOTOR - 103,088.64VA x .25 (25 percent) = 25,772.16VA

TOTAL DEMAND LOAD (LINE and NEUTRAL) (List each computed line and neutral loads below and total lines 1. - 6.)

	LINE LOAD	NEUTRAL LOAD	
		Permitted Reduction	Prohibited Reduction
1. General Lighting	104,145.00VA	--	104,145.00VA
2. Other Lighting Loads	--	--	--
3. Receptacle Loads	--	--	--
4. Other Loads	1,501,095.38VA	222,144.17VA	18,581.93VA
5. Heating and AC Equip.	--	--	--
6. Largest Motor	25,772.16VA	--	--
Total Demand Load (VA) =	1,631,012.54VA	222,144.17VA	122,726.93VA

OCCUPANCY'S OPERATING LINE VOLTAGE - $\underline{480}$V(3φ)
(Given operating voltage or as determined per test examination)

CALCULATE MINIMUM LINE and NEUTRAL LOADS
(Divide Total Demand Load [**VA**] by operating line voltage [**V**])

LINE LOAD = 1,631,012.54VA / 480V x 1.732 = 1961.86A

NEUTRAL LOAD
Permitted = 222,144.17VA / 480V x 1.732 = ~~267.21A~~* 247.05A
Prohibited = 122,726.93VA / 480V x 1.732 = 147.62A

*Where the neutral (permitted) load exceeds 200A, NEC 220.61(B)(2) permits the load to be reduced by 70 percent. Complete the following to determine the Total Neutral Load.

(1) 267.21A – 200A = 67.21A x .70 = 47.05A

(2) 47.05A + 200A = 247.05A

Total Neutral Load = 247.05A + 147.62A = 394.67A

SIZE SERVICE (Size of service based on the calculated LINE LOAD)

SIZE SERVICE REQUIRED (minimum) 2000A
(Single rating or combination - Main overcurrent device(s) to total 2000A)

Industrial Shop

98. A metal and steel industrial fabrication shop along with a 9300 square feet parts and materials storage space will be supplied by a 3-phase, 4-wire, 480/277V service. The service will supply the following loads:

277V-nonlinear

(Interior Lighting)
(57) 250W Metal Halide (MH) fixtures - 1.26A
(32) 400W Metal Halide (MH) fixtures - 1.81A
(90) 8' dual-lamp fluorescent strips - 1.67A

(Exterior Lighting)
(18) 480V dual-mounted pole fixtures - 177VA
(14) 120V wall-mounted fixtures - 1.31A

120V

(7) Copiers - 8.78A (3), 11.3A (4)
(22) Desktop computers - 1.23A
Designated receptacles (continuous) - 26
General purpose receptacles - 94
(4) 1200W microwaves
Multi-outlets (same time use) - 130'
(2) Signs - 1500VA (nonlinear)
(6) Soda machines - 12.7A (8) Vending machines - 5.6A

480V/3-phase

AC - 75,000VA
Heating - 100kW

240V/3-phase motors (10)

(5) Drill presses - 1.5HP
(3) Rotary presses - 3HP
(2) Rotary presses - 5HP

240V/1-phase motors (10)

(8) Exhaust fans - ½HP
(2) Sump pumps - 1.5HP

480V/3-phase motors (57)

(2) Air compressors - 7.5HP, 10HP
(4) Benders - 3HP (2), 5HP, 10HP
(3) Blower/Vacuums - 3HP
(5) Drill presses - 1.5HP (2), 2HP (3)
(15) Exhaust fans - 1HP
(2) Fire pumps - 40HP
(10) Grinders - ¾HP (5), 1HP (5)
(2) Jockey pumps - 3HP
(3) Metal band saws - 7.5HP
(4) Punch presses - 5HP
(2) Sump pumps 1.5HP
(3) Table saws - 2HP
(2) Water pumps - 15HP

Arc Welders, 3φ-480V (25)

Nonmotor Generator (NG) (6)
68.6A, 90% duty cycle
42.1A, 80% duty cycle (2)
27.7A, 60% duty cycle (3)
Motor Generator (MG) (9)
75.8A, 80% duty cycle (3)
54.13A, 70% duty cycle (3)
40.3A, 40% duty cycle (3)

Arc Welders, 1φ-240V (10)

Motor Generator (MG) (10)
125A, 50% duty cycle (5)
70.83A, 40% duty cycle (3)
44A, 30% duty cycle (2)

Resistance Welders, 3φ-480V (10)

37.3A, 50% duty cycle (3)
34A, 40% duty cycle (4)
21A, 25% duty cycle (3)

Resistance Welders, 1φ-240V (10)

83.33A, 50% duty cycle (3)
75A, 30% duty cycle (3)
56.96A, 15% duty cycle (4)

What size 3φ-600V switchboard is required to serve the shop? What standard size three phase transformer is needed to supply all 240V and 120V loads?

1. GENERAL LIGHTING or ACTUAL LIGHTING LOADS

General Lighting Load <NEC References - 220.12 and Table 220.12>

A. General Lighting Load - NA

Actual Lighting Load (Shop and Storage Space)

Type Fixture	VA rating		No. of Fixtures		TOTAL VA
250W MH	277V x 1.26A	x	57	=	19,894.14
400W MH	277V x 1.81A	x	32	=	16,043.84
8' Fluorescent	277V x 1.67A	x	90	=	41,633.10
					77,571.08

B. Actual Lighting Load - 77,571.08VA x 1.25 = 96,963.85VA

1.   LINE LOAD

NEUTRAL LOAD
Permitted Reduction / Prohibited Reduction

96,963.85VA   --   77,571.08VA

2. OTHER LIGHTING LOADS

A. Sign/Outline (S/O) Lighting <NEC References - 220.12(F) and 600.5(A)>

Signs (120V) - 1500VA x 2 = 3000VA

Total (Sign/Outline Lighting) = 3000VA

B. Outside Lighting <NEC Reference - 220.18(B)>

Dual-mounted pole fixtures (480V) - 177VA x 18   = 3186.0VA

Wall-mounted fixtures (120V) - 120V x 1.31A x 14   = 2200.8VA

Total (Outside Lighting)   = 5386.8VA

C. - E. NA

F. Other Lighting (LINE and NEUTRAL) Loads Total [Add lines A. and B. where applicable]

Other Lighting Loads Total = 8386.8VA
TOTAL = 8386.8VA x 1.25 = 10,483.5VA

2.	LINE LOAD	NEUTRAL LOAD	
		(277V)	
		Permitted Reduction	Prohibited Reduction
	10,483.5VA	--	8386.8VA

3. RECEPTACLE LOADS

A. Non-continuous duty <NEC References - 220.14(H), 220.14 (I), 220.44 and Table 220.44>

  (1) Fixed Multioutlet Assemblies <NEC References - 220.14(H)>

   Multi-outlets (same time use)
   130' x 180VA = 23,400VA

  (2) General Purpose Receptacles and Fixed Multioutlet Assemblies <NEC References - 220.14(H)>

   94 x 180VA = 16,920VA + 23,400VA = 40,320VA

   APPLY DEMAND FACTORS
   a. First 10,000VA (10kVA) (@100%) = 10,000VA
   b. Remainder - 30,320VA x .50 = 15,160VA
              Total = 25,160VA

B. NA

C. Continuous duty <NEC References - 220.14(I), 215.2(A)(1) or 230.42(A)>

   (120V-Designated) 26 x 180VA = 4680VA x 1.25 = 5850VA

D. Total Receptacle (NEUTRAL and LINE) Load
   LINE = 25,160VA + 5850VA = 31,010VA
   NEUTRAL = 25,160VA + 4680VA = 29,840VA

3.	LINE LOAD	NEUTRAL LOAD	
		Permitted Reduction	Prohibited Reduction
	31,010VA	29,840VA	--

4. KITCHEN EQUIPMENT - NA

5. SPECIFIC LOADS

Type Load	Calculation		
120V			
Copiers	120V x 8.78A x 3	=	3160.8VA
Copiers	120V x 11.3A x 4	=	5424.0VA
Desktop computers	120V x 1.23A x 22	=	3247.2VA
Microwaves	1200VA x 4	=	4800.0VA
Soda machines	120V x 12.7A x 6	=	9144.0VA
Vending machines	120V x 5.6A x 8	=	5376.0VA
			31,152.0VA

5.	LINE LOAD		NEUTRAL LOAD	
			Permitted Reduction	Prohibited Reduction
	31,152VA		31,152VA	--

6. MOTOR LOADS

A. Continuous Duty

Motor Load	Calculation		
240V-1ϕ			
Exhaust fans (½HP)	240V x 4.9A** x 8	=	9408.00VA
Sump pumps (1.5HP)	240V x 10A** x 2	=	4800.00VA
**Table 430.248			14,208.00VA
240V-3ϕ			
Drill presses (1.5HP)	240V x 6A* x 1.732 x 5	=	12,470.40VA
Rotary presses (3HP)	240V x 9.6A* x 1.732 x 3	=	11,971.58VA
Rotary presses (5HP)	240V x 15.2A* x 1.732 x 2	=	12,636.67VA
			37,078.65VA
480V-3ϕ			
Air compressor (7.5HP)	480V x 11A* x 1.732	=	9144.96VA
Air compressor (10HP)	480V x 14A* x 1.732	=	11,639.04VA
Benders (3HP)	480V x 4.8A* x 1.732 x 2	=	7981.06VA
Bender (5HP)	480V x 7.6A* x 1.732	=	6318.34VA
Bender (10HP)	480V x 14A* x 1.732	=	11,639.04VA
Blower/Vacuums (3HP)	480V x 4.8A* x 1.732 x 3	=	11,971.58VA
Drill presses (1.5HP)	480V x 3A* x 1.732 x 2	=	4988.16VA
Drill presses (2HP)	480V x 3.4A* x 1.732 x 3	=	8479.87VA
Exhaust fans (1HP)	480V x 2.1A* x 1.732 x 15	=	26,187.84VA

Fire pumps (40HP)	480V x 52A* x 1.732 x 2	=	86,461.44VA
Grinders (¾HP)	480V x 1.6A* x 1.732 x 5	=	6650.88VA
Grinders (1HP)	480V x 2.1A* x 1.732 x 5	=	8729.28VA
Jockey pumps (3HP)	480V x 4.8A* x 1.732 x 2	=	7981.06VA
Metal band saws (7.5HP)	480V x 11A* x 1.732 x 3	=	27,434.88VA
Punch presses (5HP)	480V x 7.6A* x 1.732 x 4	=	25,273.34VA
Sump pumps (1.5HP)	480V x 3A* x 1.732 x 2	=	4988.16VA
Table Saws (2HP)	480V x 3.4A* x 1.732 x 3	=	8479.87VA
Water pumps (15HP)	480V x 21A* x 1.732 x 2	=	34,917.12VA
*Table 430.250			309,265.92VA

B. - D. - NA

E. Total Motor Loads - [LINE LOAD - Add lines (A.) and (D.)] - [NEUTRAL LOAD - Total motor loads with neutral connections (120V)]

TOTAL =   360,552.57VA

6.   LINE LOAD

360,552.57VA

NEUTRAL LOAD

Permitted	Prohibited
Reduction	Reduction
--	--

7. MEDICAL EQUIPMENT - NA

8. INDUSTRIAL EQUIPMENT - NA

Arc Welders

(1) Individual Welders – NA

(2) Group of Welders <NEC References - 630.11(B)>

Two Largest Welders [100% (1)]
(MG-50%) 125A x .75 x 240V x 2   =   45,000.0VA

Third Largest Welder [85% (.85)]
(MG-50%) 125A x .75 x 240V x .85   =   19,125.00VA

Fourth Largest Welder [70% (.70)]
(MG-50%) 125A x .75 x 240V x .70   =   15,750.00VA

Remaining Welders [60% (.60)]
(MG-50%) 125A x .75 x 240V x .60   =   13,500.00VA
(MG-80%) 75.8A x .91 x 480V x 1.732 x .60 x 3   =   103,222.00VA
(MG-40%) 70.83A x .69 x 240V x .60 x 3   =   21,113.00VA

(NG-90%) 68.6A x .95 x 480V x 1.732 x .60	=	32,507.84VA
(MG-70%) 54.13A x .86 x 480V x 1.732 x .60 x 3	=	69,662.35VA
(MG-30%) 44A x .62 x 240V x .60 x 2	=	7856.64VA
(NG-80%) 42.1A x .89 x 480V x 1.732 x .60 x 2	=	37,380.27VA
(MG-40%) 40.3A x .69 x 480V x 1.732 x .60 x 2	=	27,741.15VA
(NG-60%) 27.7A x .78 x 480V x 1.732 x .60 x 3	=	<u>32,332.26VA</u>

(Total Arc Welders) 425,190.51VA

Resistance Welders

(1) Individual Welders – NA

(2) Group of Welders <NEC References - 630.31(B)>

Largest Welder [100% (1)]
(50%) 83.33A x .71 x 240V	=	14,199.43VA

Remaining Welders [60% (.60)]
(50%) 83.33A x .71 x 240V x .60 x 2	=	17,039.32VA
(30%) 75A x .55 x 240V x .60 x 3	=	17,820.00VA
(15%) 56.96A x .39 x 240V x .60 x 4	=	12,795.50VA
(50%) 37.3A x .71 x 480V x 1.732 x .60 x 3	=	39,630.43VA
(40%) 34A x .63 x 480V x 1.732 x .60 x 4	=	42,738.60VA
(25%) 21A x .50 x 480V x 1.732 x .60 x 3	=	<u>15,712.70VA</u>

(Total Resistance Welders) 159,935.98VA

8.   <u>LINE LOAD</u>

	<u>NETRAL LOAD</u>	
	Permitted Reduction	Prohibited Reduction
585,126.49VA	--	--

9. HEATING and AIR-CONDITIONING (AC) EQUIPMENT <NEC References - 220.50, 220.51, 220.60, 430.6(A)(1) and 440.6(A)>

ELECTRICAL HEATING UNITS
Heating Unit	Calculation
100kW (3φ)	100,000W

        TOTAL HEAT = 100,000VA (Larger)

AIR-CONDITIONING(AC) UNITS
AC Unit	Calculation
75,000VA (3φ)	75,000VA

        TOTAL AC = 75,000VA

9.   LINE LOAD   NETRAL LOAD

	Permitted Reduction	Prohibited Reduction

100,000VA   --   --

10. LARGEST MOTOR

Fire Pump (40HP -52A*) (highest current per NEC 430.17)
480V x 52A x 1.732 x .25 (25 percent) = 10,807.68VA
*Table 430.250

10.   LINE LOAD   NEUTRAL LOAD

	Permitted Reduction	Prohibited Reduction

10,807.68VA   --   --

TOTAL DEMAND LOAD (LINE and NEUTRAL) (List each computed line and neutral loads below and total lines 1. – 10.)

	LINE LOAD	NEUTRAL LOAD Permitted Reduction	Prohibited Reduction
1. General Lighting	96,963.85VA	--	77,571.08VA
2. Other Lighting Loads	10,483.50VA	--	8386.80VA
3. Receptacle Loads	31,010.00VA	29,840.0VA	--
4. Kitchen Equipment	--	--	--
5. Specific Loads	31,152.00VA	31,152.0VA	--
6. Motor Loads	360,552.57VA	--	--
7. Medical Equipment	--	--	--
8. Industrial Equipment	585,126.49VA	--	--
9. Heating and AC Equip.	100,000.00VA	--	--
10. Largest Motor	10,807.68VA	--	--
TOTAL =	1,226,096.09VA	60,992.0VA	85,957.88VA

OCCUPANCY'S OPERATING LINE VOLTAGE - 480V(3ϕ)
(Given operating voltage or as determined per test examination)

CALCULATE MINIMUM LINE and NEUTRAL LOADS
(Divide Total Demand Load **[VA]** by operating line voltage **[V]**)

LINE LOAD = 1,226,096.09VA / 480V x 1.732 = 1474.81A

SIZE SERVICE (Size of service based on the calculated **LINE LOAD**)

SIZE SERVICE (SWITCHBOARD) REQUIRED (minimum) 1600A

Transformer Sizing Tips For Closed Delta 3-Phase Units

1. Calculate all loads at 100 percent that is, no continuous or derating factors applied.
2. Round all VA ratings to nearest whole value.
3. Divide all three-phase loads (per VA ratings) by 3 and list per phase.
4. Divide all single-phase 240V loads (per VA ratings) by 2 and list per two phases.
5. List each 120V load among the two phases that will provide 120V to neutral.
6. Ensure all single-phase 240V and 120V neutral loads are evenly balance (as much as possible) per designated phases.
7. Applying format as below-vertical and horizontal line calculations should yield the same results.

Transformer for 240V and 120V Loads (B-Phase, High Leg)

Load	A	B	C	Total
Exhaust fans (8)	588VA	588VA	--	1176VA
240V-1φ	--	588VA	588VA	1176VA
	588VA	--	588VA	1176VA
	588VA	588VA	--	1176VA
	-	588VA	588VA	1176VA
	588VA	--	588VA	1176VA
	588VA	588VA	--	1176VA
	-	588VA	588VA	1176VA
Sump pumps (2)	1200VA	--	1200VA	2400VA
240V-1φ	1200VA	1200VA	--	2400VA
Arc Welders (5)	--	15,000VA	15,000VA	30,000VA
240V-1φ (125A)	15,000VA	--	15,000VA	30,000VA
	15,000VA	15,000VA	--	30,000VA
	--	15,000VA	15,000VA	30,000VA
	15,000VA	15,000VA	--	30,000VA
Arc Welders (3)	8500VA	8500VA	--	17,000VA
240V-1φ (70.83A)	--	8500VA	8500VA	17,000VA
	8500VA	--	8500VA	17,000VA
Arc Welders (2)	5280VA	5280VA	--	10,560VA
240V-1φ (44A)	--	5280VA	5280VA	10,560VA
Resist. Welders (3)	10,000VA	--	10,000VA	20,000VA
240V-1φ (83.33A)	10,000VA	10,000VA	--	20,000VA
	--	10,000VA	10,000VA	20,000VA
Resist. Welders (3)	9000VA	--	9000VA	18,000VA
240V-1φ (75A)	9000VA	9000VA	--	18,000VA
	--	9000VA	9000VA	18,000VA
Resist. Welders (4)	6835VA	--	6835VA	13,670VA
240V-1φ (56.96AA)	6835VA	6835VA	--	13,670VA
	--	6835VA	6835VA	13,670VA
	6835VA	--	6835VA	13,670VA

Drill Presses (5)	831VA	831VA	831VA	2493VA
240V-3φ (1.5HP)	831VA	831VA	831VA	2493VA
	831VA	831VA	831VA	2493VA
	831VA	831VA	831VA	2493VA
	831VA	831VA	831VA	2493VA
Rotary Presses (3)	1330VA	1330VA	1330VA	3990VA
240V-3φ (3HP)	1330VA	1330VA	1330VA	3990VA
	1330VA	1330VA	1330VA	3990VA
Rotary Presses (2)	2106VA	2106VA	2106VA	6318VA
240V-3φ (5HP)	2106VA	2106VA	2106VA	6318VA
Wall-mt fixtures (14)	1100VA	--	1100VA	2200VA
Copiers (3)	1054VA	--	1054VA	2108VA
	1054VA	--	--	1054VA
Copiers (4)	2722VA	--	2722VA	5444VA
Desktop Comp's (22)	1624VA	--	1624VA	3248VA
Receptacles (26)	2340VA	--	2340VA	4680VA
GP Receptacles (94)	8460VA	--	8460VA	16,920VA
Microwaves (4)	2400VA	--	2400VA	4800VA
Multioutlets (130')	11,700VA	--	11,700VA	23,400VA
Signs (2)	1500VA	--	1500VA	3000VA
Soda Machines (6)	4572VA	--	4572VA	9144VA
V. Machines (8)	2688VA	--	2688VA	5376VA
	184,696VA	156,315VA	182,442VA	523,453VA

Per standard sizes use either three (3) single-phase 200kVA transformers totaling 600kVA or one (1) three-phase 750kVA transformer.

Office

99. A 4170SF architectural and engineering firm's office is supplied by a 240/120, 4W, 3-phase service and supplies the following loads:

120V

Track Lighting - 45 ft.	Sign - 1680VA
Receptacles - 24 (continuous)	Multioutlet assembly - 50' (Infrequent use)
Receptacles - 73 (non-continuous)	Soda and Vending Machines - 1560VA each
Water Fountains (2) - 6.5A	Exhaust fans (2) - ¼HP
Computers (25) - 1.3A (continuous)	Printers (11) - 4.3A
Copiers (4) - 11.87A	Blue Print Copiers (2) - 1140VA
Coffeemakers (2) - 880W	Microwave - 1.8kW
Garbage Disposal - ¾HP	Refrigerator - 1167VA

240V, 1-Phase
Laminator - 1.27kW
Water Heater - 8.5kW
Sump Pump - 2HP

240V, 3-Phase
Heating - 35kW
Air Handlers (2) - 5HP
10 ton A/C - 47A

Outside Lighting - 2865VA
Copier - 5280VA

5 ton A/C - 22.4A
Fan Motors (3) - ½HP
Main Frame Computer - 37A (continuous)

What size main distribution panelboard (MDP) is needed for the office? What size service, neutral and grounding electrode conductors are required to supply the panelboard? All conductors to be copper and rated for 75°C.

1. GENERAL LIGHTING or ACTUAL LIGHTING LOADS

General Lighting Load <NEC References - 220.12 and Table 220.12>

$$4170SF \times 3.5VA = 14,595VA$$

A. General Lighting Load - 14,595VA x 1.25 = 18,243.75VA

Actual Lighting Load - NA

1.	LINE LOAD	NEUTRAL LOAD
	18,243.75VA	14,595VA

2. OTHER LIGHTING LOADS

A. Sign/Outline (S/O) Lighting <NEC References - 220.12(F) and 600.5(A)>

Sign - 1680VA (120V)

Total (Sign/Outline Lighting) = 1680VA

B. Outside Lighting <NEC Reference - 220.18(B)>

Total (Outside Lighting) = 2865VA (240V)

C. NA

D. Track Lighting <NEC Reference - 220.43(B)> (Voltage rating -120V)

$$(45' \div 2') \times 150VA = 3375VA$$

E. NA

F. Other Lighting (LINE and NEUTRAL) Loads Total [Add lines (A) - (E)]

Other Lighting Lights Total = 7920VA

TOTAL = 7920VA x 1.25 = 9900VA

2.   <u>LINE LOAD</u>   <u>NEUTRAL LOAD</u>

    9900VA   5055VA

3. RECEPTACLE LOADS

A. Non-continuous duty <NEC References - 220.14(H), 220.14(I), 220.44 and Table 220.44>

    (1) Fixed Multioutlet Assemblies <NEC References - 220.14(H)>

        Non-simultaneous use (Infrequent use) - (50' ÷ 5') x 180VA = 1800VA

    (2) General Purpose Receptacles* and Fixed Multioutlet Assemblies

        0* + 1800VA = 1800VA

B. Non-continuous duty (Office) <NEC 220.14(K)>

    (1) 73 x 180VA = 13,140VA*

        Since the results of B. (1)* is larger than B. (2)**, the results of 3.A.(2) can be combined with B. (1) followed by the application of the demand factors to obtain a smaller receptacle demand between the two non-continuous receptacle loads opposed to gathering a larger demand using two individual calculations. As a result, the total receptacle load prior to applying the demand load amounts to 14,940VA [13,140VA + 1800VA].

        APPLY DEMAND FACTORS
        a. First 10,000VA (10kVA) [@100%] = 10,000VA
        b. Remainder - 4940VA x .50   =   <u>2,470VA</u>
                      Total = 12,470VA

    (2) 4170SF x 1VA/SF = 4170VA**

C. Continuous duty <NEC References - 220.14(I), 215.2(A)(1) or 230.42(A)>

        24 x 180VA = 4320VA x 1.25 = 5400VA

D. Total Receptacle (NEUTRAL and LINE) Load
    LINE = 12,470VA + 5400VA = 17,870VA
    NEUTRAL = 12,470VA + 4320VA = 16,790VA

3.   <u>LINE LOAD</u>   <u>NEUTRAL LOAD</u>

    17,870VA   16,790VA

4. KITCHEN EQUIPMENT - NA

5. SPECIFIC LOADS

Type Load	Calculation		
120V			
Blue print copiers (2)	1140VA x 2	=	2280.0VA
Coffeemakers (2)	880VA x 2	=	1760.0VA
Computers (25)	120V x 1.3A x 25 x 1.25	=	4875.0VA
Copiers (4)	120V x 11.87A x 4	=	5697.6VA
Microwave		=	1800.0VA
Printers (11)	120V x 4.3A x 11	=	5676.0VA
Refrigerator		=	1167.0VA
Soda and Vending Machines	1560VA x 2	=	3120.0VA
Water fountains (2)	120V x 6.5A x 2	=	1560.0VA
240V-1ϕ			
Copier		=	5280.0VA
Laminator		=	1270.0VA
Water Heater		=	8500.0VA
240V-3ϕ			
Main Frame Computer	240V x 37A x 1.732 x 1.25	=	19,225.2VA

TOTAL = 62,210.8VA

5.	LINE LOAD	NEUTRAL LOAD (120V)
	62,210.8VA	27,935.6VA

6. MOTOR LOADS

A. Continuous Duty

Motor Load	Calculation
120V ¼HP exhaust fans (2)	120V x 5.8A* x 2 = 1392VA
240V 2HP sump pump	240V x 12A* = 2880VA

B. Non-Continuous Duty

Motor Load	Calculation
120V ¾HP garbage disposal	120V x 13.8A* = 1656VA

*Table 430.248

C. and D. NA

E. Total Motor Loads - [LINE LOAD - Add lines A. and B.] - [NEUTRAL LOAD - Total motor loads with neutral connections (120V)]

$$\text{TOTAL} = 4488\text{VA}$$

6. <u>LINE LOAD</u>   <u>NEUTRAL LOAD</u>
                                                     (120V)

     5928VA   3048VA

7. MEDICAL EQUIPMENT - NA

8. INDUSTRIAL EQUIPMENT - NA

9. HEATING and AIR-CONDITIONING (AC) EQUIPMENT <NEC References - 220.50, 220.51, 220.60, 430.6(A)(1) and 440.6(A)>

ELECTRICAL HEATING UNITS

Heating Unit	Calculation
35kW (3ϕ)	35,000W

          TOTAL HEAT = 35,000VA

AIR-CONDITIONING(AC) UNITS

AC Unit	Calculation
10 ton AC/Fan mtrs (2)	240V x 47A + (240V x 2.2A* x 2) x 1.732 = 21,365.95VA
5 ton AC/Fan mtr	240V x 22.4A + (240V x 2.2A*) x 1.732 = <u>10,225.73VA</u>

                               TOTAL AC = 31,591.68VA

Line Load =   35,000VA   +   (12,636.67VA) = 47,636.67VA
               (Largest Load)   (Air Handlers)

         Air Handlers - 240V x 15.2A* x 2 x 1.732 = 12,636.67VA
         *Table 430.250

9. <u>LINE LOAD</u>   <u>NETRAL LOAD</u>

     47,636.67VA   0

10. LARGEST MOTOR - Air Handler - 240V x 15.2A x 1.732 x .25(25 percent) = 1579.6VA

10. <u>LINE LOAD</u>   <u>NEUTRAL LOAD</u>

     1579.6VA   0

TOTAL DEMAND LOAD (LINE and NEUTRAL) (List each computed line and neutral loads below and total lines 1. - 10.)

	LINE LOAD	NEUTRAL LOAD
1. General Lighting	18,243.75VA	14,595.0VA
2. Other Lighting Loads	9900.00VA	5055.0VA
3. Receptacle Loads	17,870.00VA	16,790.0VA
4. Kitchen Equipment	--	--
5. Specific Loads	62,210.80VA	27,935.6VA
6. Motor Loads	5928.00VA	3048.0VA
7. Medical Equipment	--	--
8. Industrial Equipment	--	--
9. Heating and AC Equipment	47,636.60VA	0
10. Largest Motor	1579.60VA	0
TOTAL =	163,368.75VA	67,423.6VA

OCCUPANCY'S OPERATING LINE VOLTAGE - 240V(3φ)
(Given operating voltage or as determined per test examination)

CALCULATE MINIMUM LINE AND NEUTRAL LOADS
(Divide Total Demand Load [VA] by operating line voltage [V])

LINE LOAD = 163,368.75VA / 240V x 1.732 = 393.02A
NEUTRAL LOAD = 67,423.6VA / 240V x 1.732 = 162.20A

SIZE SERVICE (Size of service based on the calculated **LINE LOAD**)

SIZE SERVICE REQUIRED (minimum) 400A
(Single rating or combination - Main overcurrent device(s) to total **400A**)

SIZING FEEDER/SERVICE CONDUCTORS <NEC References - 215.2(A), 230.42(A), 240.4(B) and Table 310.15(B)(16)> (Based on the calculated Line Loads and NEC References)

Per Table 310.15(B)(16) at 75°C the minimum size copper conductors required to supply a 400A service are 500 kcmil which has a rated ampacity of 380 amps. Although the ampacity of the conductors is less than the 400A service they will supply, the use of these conductors are permitted per NEC 240.4(B).

FEEDER/SERVICE CONDUCTORS 500 kcmil copper

SIZING NEUTRAL CONDUCTOR <NEC References - 215.2(A)(2), 220.61, 230.42(C), 250.24(C) and Table 310.15(B)(16)> (Based on the calculated Neutral Load and NEC References)

Per Table 310.15(B)(16) at 75°C the minimum size copper conductor required to serve as the neutral conductor is a 2/0 AWG which has a rated ampacity of 175 amps.

NEUTRAL CONDUCTOR(S) 2/0 AWG copper

SIZING GROUNDING ELECTRODE CONDUCTOR <NEC References - 250.24(C)(1), Table 250.66 and Table 8 of Chapter 9>

Per Table 250.66 based on the use of 500 kcmil copper service conductors the grounding electrode conductor must be a 1/0 AWG copper which is smaller than the 2/0 AWG copper neutral conductor.

GROUNDING ELECTRODE CONDUCTOR 1/0 AWG copper

Restaurant

100. A new all electric 5000SF restaurant will be supplied by a 3-phase, 208/120V service. The restaurant will be equipped with the following loads:

120V

Menu signs lighting - 2345VA
Beverage dispensers (2) - 10.6A
Receptacles (10) (continuous)
Outside parking lights (13) - 2.8A
Cash registers (2) - 3.7A

100' track lighting
Exhaust fans (2) - ¼HP (continuous)
Receptacles (30) (noncontinuous)
1800VA sign

Kitchen Equipment

2 - 1680VA refrigerators
2 - 1400VA bun warmers
2 - 1600VA mixers

2 - 1.9kW coffeemakers
2 - 1600VA toasters
1 - Ice cream box - 13.87A

208V - 3 Phase

AC - 19,400VA
Heating - 35kW

Kitchen Equipment

2 - 6kW deep fryers
1 - 3450VA mixer
2 - 14kW cooktop grills
3 - 10kW ovens
1 - 6.5kVA freezer
1 - 7kW booster heater

1 - 10kW broiler
2 - 1HP exhaust fans
1 - 3HP grinder
1 - 4.5kVA walk-in cooler
1 - 3.8kW dishwasher
2 - 4760VA meat/poultry freezer

Determine the load using the standard calculation. What size service is required? All lighting loads are inductive (nonlinear).

1. GENERAL LIGHTING or ACTUAL LIGHTING LOADS

General Lighting Load <NEC References - 220.12 and Table 220.12>

Restaurant - 5000SF x 2VA = 10,000VA

A. General Lighting Load - 10,000VA x 1.25 = 12,500VA

Actual Lighting Load - NA

1.	LINE LOAD	NEUTRAL LOAD	
		Permitted Reduction	Prohibited Reduction
	12,500VA	--	10,000VA

2. OTHER LIGHTING LOADS

A. Sign/Outline (S/O) Lighting <NEC References - 220.12(F) and 600.5(A)>

Sign (120V) = 1800VA

Total (Sign/Outline Lighting) = 1800VA

B. Outside Lighting <NEC Reference - 220.18(B)>

Outside parking lights - 120V x 2.8A x 13 = 4368.2VA

Total (Outside Lighting) = 4368.2VA

C. NA

D. Track Lighting <NEC Reference - 220.43(B)> (Voltage rating -120V)

(100' ÷ 2') x 150VA = 7500VA

E. Miscellaneous (Write-ins. List individual voltage rating of each lighting load)

Menu signs lighting (120V) = 2345VA

Total (Miscellaneous) = 2345VA

F. Other Lighting (LINE and NEUTRAL) Loads Total [Add lines A. - E. where applicable]

Other Lighting Loads Total = 16,013.2VA

TOTAL = 16,013.2VA x 1.25 = 20,016.5VA

2. LINE LOAD

NEUTRAL LOAD

(120V)

Permitted Reduction	Prohibited Reduction

20,016.5VA

-- 16,013.2VA

3. RECEPTACLE LOADS

A(1) - NA

A(2) Non-continuous duty - General Purpose Receptacles <NEC 220.14(I)>

30 x 180VA = 5400VA* (Demand factors not applicable)

B. NA

C. Continuous duty <NEC References - 220.14(I), 215.2(A)(1) or 230.42(A)>

10 x 180VA = 1800VA* x 1.25 = 2250VA

D. Total Receptacle (LINE and NEUTRAL*) Load

Total Receptacle loads = 7650VA

3. LINE LOAD

NEUTRAL LOAD

Permitted Reduction	Prohibited Reduction

7650VA

7200VA --

4. KITCHEN EQUIPMENT (120V-underlined)

Type Kitchen	Calculation		
Booster heater		=	7000.00VA
Broiler		=	10,000.00VA
Bun warmers	1400VA x 2	=	2800.00VA
Cooktop grills	14,000VA x 2	=	28,000.00VA
Coffee makers	1900VA x 2	=	3800.00VA
Deep Fryers	6000VA x 2	=	12,000.00VA
Dishwasher		=	3800.00VA
Freezer		=	6500.00VA

Grinder	208V x 10.6A* x 1.732	=	3818.71VA
<u>Ice cream box</u>	120V x 13.87A	=	1664.40VA
Meat/poultry freezer	4760VA x 2	=	9520.00VA
Mixer (3φ)		=	3450.00VA
<u>Mixers</u>	1600VA x 2	=	3200.00VA
Ovens	10,000VA x 3	=	30,000.00VA
<u>Refrigerators</u>	1680VA x 2	=	3360.00VA
<u>Toasters</u>	1600VA x 2	=	3200.00VA
Walk-in cooler		=	<u>4500.00VA</u>

*Table 430.250

(29) Total = 136,613.11VA

a. Per Table 220.56 - 136,613.11VA x .65 = 88,798.52VA
b. Two Largest Kitchen Loads - Cooktop grills-14,000VA x 2 = 28,000.00VA

KITCHEN LOAD (Larger of a. and b.) = 88,798.52VA

NEUTRAL LOAD (120V) (11) - 18,024.4VA x .65 = 11,715.86VA

4.	LINE LOAD	NEUTRAL LOAD	
		Permitted Reduction	Prohibited Reduction
	88,798.52VA	11,715.86VA	--

5. SPECIFIC LOADS

Load	**Calculation**		
Beverage dispensers	120V x 10.6A x 2	=	2544VA
Cash registers	120V x 3.7A x 2	=	<u>888VA</u>
		TOTAL =	3432VA

5.	LINE LOAD	NEUTRAL LOAD	
		Permitted Reduction	Prohibited Reduction
	3432VA	3432VA	--

6. MOTOR LOADS

A. Continuous Duty

Motor Load	Calculation	
120V Exhaust fans (¼ HP)	120V x 5.8A* x 2	= 1392VA
*Table 430.248		
208V Exhaust fans (1 HP)	208V x 4.6A** x 2 x 1.732	= 3314.36VA
**Table 430.250		

B. - D. NA

E. Total Motor Loads - [LINE LOAD - Add line A.] - [NEUTRAL LOAD - Total motor loads with neutral connections (120V)]

TOTAL = 4706.36VA

6.	LINE LOAD	NEUTRAL LOAD	
		Permitted Reduction	Prohibited Reduction
	4706.36VA	1392VA	--

7. MEDICAL EQUIPMENT - NA

8. INDUSTRIAL EQUIPMENT - NA

9. HEATING and **AIR-CONDITIONING (AC) EQUIPMENT** <NEC References - 220.50, 220.51, 220.60, 430.6(A)(1) and 440.6(A)>

HEAT = 35kW
AC = 19,400VA
Line Load = 35,000VA (Larger)

9.	LINE LOAD	NEUTRAL LOAD	
		Permitted Reduction	Prohibited Reduction
	35,000VA	0	--

10. LARGEST MOTOR

Mixer (1600VA/120V = 13.33A) (highest current per NEC 430.17) although there are two mixers with similar ratings only one is considered. Even though the mixer is considered kitchen equipment it's still operated by a motor which in this situation has the highest current rating.

1600VA x .25 (25 percent) = 400VA

10.	LINE LOAD	NEUTRAL LOAD	
		Permitted Reduction	Prohibited Reduction
	400VA	400VA	--

TOTAL DEMAND LOAD (LINE and NEUTRAL) (List each computed line and neutral loads below and total lines 1. – 10.)

	LINE LOAD	NEUTRAL LOAD Permitted Reduction	NEUTRAL LOAD Prohibited Reduction
1. General Lighting	12,500.00VA	--	10,000.0VA
2. Other Lighting Loads	20,016.50VA	--	16,013.2VA
3. Receptacle Loads	7650.00VA	7200.00VA	--
4. Kitchen Equipment	88,798.52VA	11,715.86VA	--
5. Specific Loads	3432.00VA	3432.00VA	--
6. Motor Loads	4706.36VA	1392.00VA	--
7. Medical Equipment	--	--	--
8. Industrial Equipment	--	--	--
9. Heating and AC Equip.	35,000.00VA	0	--
10. Largest Motor	400.00VA	400.00VA	--
TOTAL =	172,503.38VA	24,139.86VA	26,013.2VA

OCCUPANCY'S OPERATING LINE VOLTAGE - 208V(3ϕ)
(Given operating voltage or as determined per test examination)

CALCULATE MINIMUM LINE and NEUTRAL LOADS
(Divide Total Demand Load **[VA]** by operating line voltage **[V]**)

LINE LOAD = 172,503.38VA / 208V x 1.732 = 478.84A

NEUTRAL LOAD
     Permitted = 24,139.86VA / 208V x 1.732 = 67.01A
     Prohibited = 26,013.2VA / 208V x 1.732 = 72.21A

Total Neutral Load = 67.01A + 72.21A = 139.22A

SIZE SERVICE (Size of service based on the calculated LINE LOAD)

SIZE SERVICE REQUIRED (minimum) 500A
(Single rating or combination - Main overcurrent device(s) to total 500A)

See question No. 127. (NEC 220.88 - Optional Calculation for Restaurants)

School

101. A community college totaling 495,000SF in its entirety will be supplied by a 4160/2400V, 3ϕ, 4-wire system with a distribution voltage at 480/277V. The general lighting load for this occupancy is all rated for 277V. Over 80 percent of the lighting is HID-rated. Use the accompanying information to perform a standard load calculation and to determine the service and neutral loads for the entire community college.

120V
Receptacles - 23,488 (non-continuous)
Receptacles - 1306 (continuous)
Multioutlets - 296ft. (HD)
Multioutlets - 735ft. (LD)
Soda Machines (77) - 6.5A
Snack Machines (58) - 5.2A
Water fountains (64) - 4.8A
Desktop computers (3567) - 1.38A
Laser printers (310) - 1.3A
Exhaust fans (206) - ¼HP
Vending microwaves (42) - 1000W
Commercial washers (20) - ¾HP

Kitchen Equipment
Coffeemakers (14) - 1540W
Refrigerators (13) - 1800VA
Microwave ovens (13) - 2000W
Vegetable peelers (10) - 2275VA
Toasters (25) - 1340VA
Bun warmers (16) - 1.35kW

277V
Metal Halide wall-mtd fixtures (595) - 2.7A
Show Window - 1120ft.
Signs (5) - 2450VA
Track Lighting - 1464ft.
60 - receptacles (continuous-switched controlled lighting)
Other outside lighting - 76,000VA

208/120V-1φ*
20 - 10kW ranges (instructional)
Copiers (22) - 18.8A (continuous)
Commercial dryers (17) - 7.8kW

*Neutral loads of 208/120V-1φ equipment @ 20 percent of line loads

208V-3φ
Main frame computers (21) - 22.6kVA (continuous)
Copiers (46) - 15.7A (continuous)

208V-1φ - Kitchen Equipment
12kW deep fryers (4)   10kW broilers (3)   Mixers (2) - 3450VA   Freezers (3) - 3970VA
Dishwashers (3) - 1kW   2.7kW meat grinders (2)
Instant heaters (3) - 4kW   Ice Cream Freezers (3) - 1545VA

480V-1φ
Outside pole lighting (368) - 400W fixtures (2.1A each)

480V-3φ
Heating - 967kW   100 tons AC units (45) - 103A   Air Handlers (67) - 22.44A
120 tons AC units (31) - 122A   AC fan motors (152) - 11.6A   Fire water pumps (16) - 75HP
Jockey pumps (16) - 5HP   Elevator motors (22) - 60HP   Elevator motors (16) - 60HP
Intermittent duty/continuous   Intermittent duty/15-minute   Exhaust fans (16) - 1HP
Return Air motors (28) - 100HP   Water pumps (18) - 50HP
Escalator motors (8) - 60HP (continuous)   Water Heaters (27) - 13.5kW

Kitchen Equipment
14kW Grill cooktops (20)   Ovens (13) - 9.5kW
Ovens (15) - 15.5kW   Water Heaters (14) - 7.5kW

1. GENERAL LIGHTING or ACTUAL LIGHTING LOADS

General Lighting Load <NEC References - 220.12 and Table 220.12>

School - 495,000SF x 3VA = 1,485,000VA

A. General Lighting Load - 1,485,000VA x 1.25 = 1,856,250VA

Actual Lighting Load - NA

1.	LINE LOAD	NEUTRAL LOAD	
		Permitted Reduction	Prohibited Reduction
	1,856,250VA	--	1,485,000VA

2. OTHER LIGHTING LOADS

A. Sign/Outline (S/O) Lighting <NEC References - 220.12(F) and 600.5(A)>

Signs (277V) - 2450VA x 5 = 12,250VA
Total (Sign/Outline Lighting) = 12,250VA

B. Outside Lighting <NEC Reference - 220.18(B)>

Other outside lighting (277V) -   =   76,000.0VA
Outside pole lighting - 480V x 2.1A x 368 =   370,944.0VA
Total (Outside Lighting) =   446,944.0VA

C. Show-Window Lighting <NEC Reference - 220.43(A)> (Voltage rating - 277V)

1120' x 200VA = 224,000VA

D. Track Lighting <NEC Reference - 220.43(B)> (Voltage rating - 277V)

(1464' ÷ 2') x 150VA = 109,800VA

E. Miscellaneous

Metal Halide wall-mtd fixtures - 277V x 2.7A x 595 = 445,000.5VA

Total (Miscellaneous) = 445,000.5VA

F. Other Lighting (LINE and NEUTRAL) Loads Total [Add lines A. - E. where applicable]

Other Lighting Loads Total = 1,237,994.5VA

TOTAL = 1,237,994.5VA x 1.25 = 1,547,493.13VA

2.	LINE LOAD	NEUTRAL LOAD

<table>
<tr><td></td><td></td><td colspan="2" align="center">(277V)</td></tr>
<tr><td></td><td></td><td>Permitted
Reduction</td><td>Prohibited
Reduction</td></tr>
<tr><td></td><td>1,547,493.13VA</td><td>--</td><td>867,050.5VA</td></tr>
</table>

3. RECEPTACLE LOADS

A. Non-continuous duty <NEC References - 220.14(H) and (I), 220.44 and Table 220.44> (120V)

(1) Fixed Multioutlet Assemblies <NEC References - 220.14(H)>

Non-simultaneous use (Plugmolds-light duty) - (735' ÷ 5') x 180VA = 26,460VA

Simultaneous use (Plugmolds-heavy duty) - 296' x 180VA = 53,280VA

(2) General Purpose Receptacles* and Fixed Multioutlet Assemblies <NEC References – 220.14(H)>

23,488* x 180VA = 4,227,840VA + 26,460VA + 53,280VA = 4,307,580VA

APPLY DEMAND FACTORS
a. First 10,000VA (10kVA) (@100%) = 10,000VA
b. Remainder - 4,297,580VA x .50 = 2,148,790VA
Total = 2,158,790VA

B. NA

C. Continuous duty <NEC References - 220.14(I), 215.2(A)(1) or 230.42(A)>

(120V) 1306 x 180VA = 235,080VA x 1.25 = 293,850VA
(277V) 60 x 180VA = 10,800VA x 1.25 = 13,500VA
307,350VA

D. Total Receptacle (NEUTRAL and LINE) Load
LINE = 2,158,790VA + 307,350VA = 2,466,140VA
NEUTRAL = 2,158,790VA + 235,080VA +10,800* = 2,404,670VA

3.	LINE LOAD	NEUTRAL LOAD

<table>
<tr><td></td><td></td><td>Permitted
Reduction</td><td>Prohibited
Reduction</td></tr>
<tr><td></td><td>2,466,140VA</td><td>2,393,870VA</td><td>10,800VA*</td></tr>
</table>

*Because the 277V switched controlled receptacles will be supplying dedicated lighting loads that could be HID rated, reduction about the neutral is prohibited.

4. KITCHEN EQUIPMENT

Type Kitchen	Calculation		

120V (91)

Bun warmers	1350VA x 16	=	21,600VA
Coffeemakers	1540VA x 14	=	21,560VA
Microwave ovens	2000VA x 13	=	26,000VA
Refrigerators	1800VA x 13	=	23,400VA
Toasters	1340VA x 25	=	33,500VA
Vegetable peelers	2275VA x 10	=	22,750VA
			148,810VA

208V-1φ (23)

Broilers	10,000VA x 3	=	30,000VA
Deep fryers	12,000VA x 4	=	48,000VA
Dishwashers	1000VA x 3	=	3000VA
Freezers	3970VA x 3	=	11,910VA
Meat grinders	2700VA x 2	=	5400VA
Ice cream freezers	1545VA x 3	=	4635VA
Instant heaters	4000VA x 3	=	12,000VA
Mixers	3450VA x 2	=	6900VA
			121,845VA

480V-3φ (62)

Grill cooktops	14,000VA x 20	=	280,000VA
Ovens	15,500VA x 15	=	232,500VA
Ovens	9500VA x 13	=	123,500VA
Water Heaters	7500VA x 14	=	105,000VA
			741,000VA

(176) Total	=	1,011,655VA

a. Per Table 220.56 (85) - 1,011,655VA x .65 = 657,575.75VA
b. Two Largest Kitchen Loads - Ovens -15,500VA x 2 = 31,000.00VA
KITCHEN LOAD (Larger of a. and b.) = 657,575.75VA

NEUTRAL LOAD (120V) (91) - 148,810VA x .65 = 96,726.5VA

4.	LINE LOAD	NEUTRAL LOAD	
		Permitted Reduction	Prohibited Reduction
	657,575.75VA	96,726.5VA	--

5. SPECIFIC LOADS

Type Load	Calculation		
120V			
Commercial washers	120V x 13.8A* x 20	=	33,120.0VA
Desktop computers	120V x 1.38A x 3567	=	590,695.2VA
Exhaust fans	120V x 5.8A* x 206	=	143,376.0VA
Laser printers	120V x 1.3A x 310	=	48,360.0VA
Snack machines	120V x 5.2A x 58	=	36,192.0VA
Soda machines	120V x 6.5A x 77	=	60,060.0VA
Vending microwaves	1000VA x 42	=	42,000.0VA
Water fountains	120V x 4.8A x 64	=	36,864.0VA
			990,667.2VA

208/120V-1φ			
Copiers	208V x 18.8A x 22 x 1.25	=	107,536.0VA
Commercial dryers	7800VA x 17	=	132,600.0VA
Ranges (instructional)	20 units (Table 220.55, Note 5)	=	35,000.0VA
			275,136.0VA[++]

208V-3φ			
Copiers	208V x 15.7A x 1.732 x 46 x 1.25	=	325,221.1VA
Main frame computers	22,600VA x 21 x 1.25	=	593,250.0VA
			918,471.1VA

480V-3φ			
Water Heaters	13,500VA x 27	=	364,500.0VA

*Table 430.248   TOTAL   = 2,548,774.3VA

5.   <u>LINE LOAD</u>   <u>NEUTRAL LOAD</u>
                                        (120V[+] and 208/120V[++])
                                        Permitted   Prohibited
                                        Reduction   Reduction

      2,548,774.3VA   990,667.2VA[+] 55,027.2VA[++]
                                        (208/120V loads 20 percent of line loads.)

6. MOTOR LOADS

A. Continuous Duty

Motor Load	Calculation		
Fire pumps (75HP)	480V x 96A* x 1.732 x 16	=	1,276,969.00VA
Escalators (60HP)	480V x 77A* x 1.732 x 8	=	512,117.80VA
Exhaust fans (1HP)	480V x 2.1A* x 1.732 x 16	=	27,933.70VA
Jockey pumps (5HP)	480V x 7.6A* x 1.732 x 16	=	101,093.38VA
Water pumps (50HP)	480V x 65A* x 1.732 x 18	=	972,691.20VA
*Table 430.250			2,890,805.08VA

B. and C. NA

D. Elevators - Intermittent Duty Cycle (38) <Table 430.22(E)>

   60HP motors-Intermit. dty/cont. - 480V x 77A x 1.732 x 22 x 1.40 =   1,971,653.4VA
   60HP motors-Intermit. dty/15min. - 480V x 77A x 1.732 x 16 x .85 =   ~~870,600.2VA~~
                                                                        ~~2,842,253.6VA~~

   2,842,253.6VA x .72** = 2,046,422.59VA (Demand Load)
   **Apply Demand Factor (Table 620.14) for Elevators (38 @ .72) (assumed - not under constant load)

E. Total Motor Loads - [LINE LOAD - Add lines A. and D.] - [NEUTRAL LOAD - Total motor loads
   with neutral connections (120V)]

                                        TOTAL =   5,733,058.68VA

6.   LINE LOAD   NEUTRAL LOAD
                                             Permitted   Prohibited
                                             Reduction   Reduction
      4,937,227.67VA   --   --

7. MEDICAL EQUIPMENT - NA

8. INDUSTRIAL EQUIPMENT - NA

9. HEATING and AIR-CONDITIONING (AC) EQUIPMENT <NEC References - 220.50,
   220.51, 220.60, 430.6(A)(1) and 440.6(A)>

ELECTRICAL HEATING UNITS

Heating Unit	Calculation
967kW (3ϕ)	967,000W
TOTAL HEAT =	967,000VA

AIR-CONDITIONING (AC) UNITS

AC Unit	Calculation		
100 tons AC	480V x 103A x 1.732 x 45	=	3,853,353.60VA
120 tons AC	480V x 122A x 1.732 x 31	=	3,144,203.52VA
Fan motors	480V x 11.6A x 1.732 x 152	=	1,465,853.95VA
	TOTAL AC	=	8,463,411.10VA

Line Load = 8,463,411.1VA + (1,249,933.13VA) + (2,886,481.92VA) = 12,599,826.2VA
             (Largest Load)   (Air Handlers)   (Return Air Motors)

      Air Handlers - 480V x 22.44A x 1.732 x 67 = 1,249,933.13VA
      Return Air Motors (100HP) - 480V x 124A* x 1.732 x 28 = 2,886,481.92VA
      *Table 430.250

9. <u>LINE LOAD</u>   <u>NETRAL LOAD</u>

	Permitted Reduction	Prohibited Reduction
12,599,826.2VA	--	--

10. LARGEST MOTOR

Return Air motor (100HP -124A*) (highest current per NEC 430.17)
480V x 124A x 1.732 x .25 (25 percent) = 25,772.2VA
*Table 430.250

10. <u>LINE LOAD</u>   <u>NEUTRAL LOAD</u>

	Permitted Reduction	Prohibited Reduction
25,772.2VA	--	--

TOTAL DEMAND LOAD (LINE and **NEUTRAL)** (List each computed line and neutral loads below and total lines 1. – 10.)

	<u>LINE LOAD</u>	<u>NEUTRAL LOAD</u>	
		Permitted Reduction	Prohibited Reduction
1. General Lighting	1,856,250.00VA	--	1,485,000.0VA
2. Other Lighting Loads	1,547,493.13VA	--	867,050.5VA
3. Receptacle Loads	2,466,140.00VA	2,393,870VA	10,800.0VA
4. Kitchen Equipment	657,575.75VA	96,726.5VA	--
5. Specific Loads	2,548,774.30VA	990,667.2VA	55,027.2VA
6. Motor Loads	4,937,227.67VA	--	--
7. Medical Equipment	--	--	--
8. Industrial Equipment	--	--	--
9. Heating and AC Equip.	12,599,826.20VA	--	--
10. Largest Motor	25,772.20VA	--	--
TOTAL =	26,639,059.25VA	3,481,263.7VA	2,417,877.7VA

Based on the standard load calculation for the entire community college the line load amounts to 26,639,059.25VA which yields a **service load** of **3697.24A** (26,639,059.25VA / 4160V x 1.732).

Combined, the **neutral load** amounts to a <u>permitted reduction</u> of **398.22A** [3,481,263.7VA /4160V x 1.732 = 483.17A - 200A = 283.17A x .70 = 198.22A + 200A = 398.22A] per NEC 220.61(B)(2) and a <u>prohibited reduction</u> of 335.58A [2,417,877.7VA /4160V x 1.732 = 335.58A] per NEC 220.61(C) totaling **733.80A** [398.22A + 335.58A].

See question No. 123. (NEC 220.86 - Optional Calculation for Schools).

Store

102. A 120' x 100' grocery store with a bakery, eatery and meat deli will be supplied by a 4W-3 phase, 240/120V service. The following loads are included:

120V

Computerized Registers - (8) 1.27A

Receptacles - 80 (non-continuous)

Receptacles - 40 (continuous)

Lighting Track - 150 ft.

Meat Slicers - ¼ HP (2)

Meat Cutters - ¾ HP (2)

Soda Dispenser - 12.8A

Dishwasher - 6.3kW

Plugmolds - 100 ft. (simultaneous)

Soda Machines (4) - 1560VA

Water Fountains (2) - 6.78A

Exhaust fans (4) - 1.75A

Signs (2) - 13.75A

Copier - 11.87A

Toasters (2) - 5.6A

Coffeemakers - 1340VA(2)

240V, 3-Phase

Air Handlers (2) - 7.5HP and 10HP

Boxed Items/Vegetables Freezers (6) - 10,000VA

Deep Fryers (2) - 5.8kW and 7.5kW

Fire Pump - 40HP

Heating - 100kW

Ice Cream Freezers (2) - 7500VA

Jockey Pump - 3HP

Meat/Variety Foods Coolers (4) - 15,000VA

Milk/Juice/Dairy Coolers (3) - 11,750VA

AC Rooftop Units (RTU) (5) - 14,500VA

AC Fan motors (5) - 1HP

Walk-In Cooler - 12,000VA

Water pumps (2) - 10HP (Intermittent duty/15mins.)

Cooktop grill - 7.67kW

240V, 1-Phase

Freezer - 13.54A

Cake/Dough Dual Mixer - 17A

Dishwasher (2) - 3kVA

Outside Lighting (22) - 2.43A

Refrigerator – 14.7A

Water Heaters (2) - 8000VA

Convection ovens (3) - 8.8kW

What size service is required to serve the grocery store? Using 75°C aluminum conductors, what size service, neutral and grounding electrode conductors are needed to supply the service? Consider the use of 3 parallel conductors per line and neutral.

1. GENERAL LIGHTING or ACTUAL LIGHTING LOADS

General Lighting Load <NEC References - 220.12 and Table 220.12>

$$120' \times 100' \times 3VA = 36,000VA$$

A. General Lighting Load - 36,000VA x 1.25 = 45,000VA

Actual Lighting Load - NA

1. <u>LINE LOAD</u>   <u>NEUTRAL LOAD</u>

45,000VA   36,000VA

2. OTHER LIGHTING LOADS

A. Sign/Outline (S/O) Lighting <NEC References - 220.12(F) and 600.5(A)>

Signs - 120V x 13.75A x 2 = 3300VA
Total (Sign/Outline Lighting) = 3300VA

B. Outside Lighting <NEC Reference - 220.18(B)>

Outside lighting - 240V x 2.43A x 22 =   12,830.4VA
Total (Outside Lighting) =   12,830.4VA

C. NA

D. Track Lighting <NEC Reference - 220.43(B)> (Voltage rating -120V)

$$150' \div 2' \times 150VA = 11,250VA$$

E. NA

F. Other Lighting (LINE and NEUTRAL) Loads Total [Add lines A. – E. where applicable]

Other Lighting Lights Total = 27,380.4VA

TOTAL = 27,380.4VA x 1.25 = 34,225.5VA

2. <u>LINE LOAD</u>   <u>NEUTRAL LOAD</u>

34,225.5VA   14,550VA

3. RECEPTACLE LOADS

A. Non-continuous duty <NEC References - 220.14(H), 220.14 (I), 220.44 and Table 220.44> (120V)

   (1) Fixed Multioutlet Assemblies <NEC References - 220.14(H)>

     Simultaneous use (Plugmolds-heavy duty)

     100' x 180VA = 18,000VA

   (2) General Purpose Receptacles* and Fixed Multioutlet Assemblies <NEC References – 220.14(H)>

     80* x 180VA = 14,400VA + 18,000VA = 32,400VA

     APPLY DEMAND FACTORS
     a. First 10,000VA (10kVA) (@100%) = 10,000VA
     b. Remainder - 22,400VA x .50 = 11,200VA
     Total = 21,200VA

B. NA

C. Continuous duty <NEC References - 220.14(I), 215.2(A)(1) or 230.42(A)>

   40 x 180VA = 7200VA x 1.25 = 9000VA

D. Total Receptacle (NEUTRAL and LINE) Load
LINE = 21,200VA + 9000VA = 30,200VA
NEUTRAL = 21,200VA + 7200VA = 28,400VA

3.   **LINE LOAD**   **NEUTRAL LOAD**

   30,200VA   28,400VA

4. KITCHEN EQUIPMENT

Type Kitchen   Calculation

120V (9)

Coffeemakers	1340VA x 2	=	2680.0VA
Dishwasher		=	6300.0VA
¾ HP Meat Cutters	120V x 13.8A* x 2	=	3312.0VA
¼ HP Meat Splicers	120V x 5.8A* x 2	=	1392.0VA
Toasters	120V x 5.6A x 2	=	1344.0VA
			15,028.0VA

240V-1φ (8)

Cake/Dough Dual Mixer	240V x 17A	=	4080.0VA
Convection ovens	8800VA x 3	=	26,400.0VA
Dishwashers	3000VA x 2	=	6000.0VA
Freezer	240V x 13.54A	=	3249.6VA
Refrigerator	240V x 14.7A	=	3528.0VA
			43,257.6VA

240V-3φ (4)

Cooktop grill		=	7670.0VA
Deep fryers (2)	5800VA + 7500VA	=	13,300.0VA
Walk-In Cooler		=	2,000.0VA
			32,970.0VA

(21) Total		=	91,255.6VA

a. Per Table 220.56 - 91,255.6VA x .65 = 59,316.14VA

b. Two Largest Kitchen Loads - (1) Oven 8.8kVA + Cooler 12 kVA = 20.8kVA (20,800VA)

KITCHEN LOAD (Larger of **a.** and **b.**) = 59,316.14VA

NEUTRAL LOAD (120V) (9) - 15,028.0VA x .65 = 9768.2VA

4.

LINE LOAD	NEUTRAL LOAD
59,316.14VA	9768.2VA

5. SPECIFIC LOADS

Type Load	Calculation		
120V			
Computerized Registers	120V x 1.27A x 8 x 1.25	=	1524.0VA
Copier	120V x 11.87A	=	1424.4VA
Soda Dispenser	120V x 12.8A	=	1536.0VA
Soda Machines	1560VA x 4	=	6240.0VA
Water fountains	120V x 6.78A x 2	=	1627.2VA
			12,351.6VA
240V-1φ			
Water Heaters	8000VA x 2	=	16,000.0VA
240V-3φ			
Boxed Items/Vegetable Frzs.	10,000VA x 6	=	60,000.0VA
Ice Cream Freezers	7500VA x 2	=	15,000.0VA
Milk/Juice/Dairy Coolers	11,750VA x 3	=	35,250.0VA
			107,250.0VA

SPECIFIC LOADS = 135,601.6VA

5.   <u>LINE LOAD</u>   <u>NEUTRAL LOAD</u>
                                                          (120V)
      135,601.6VA   12,351.6VA

6. MOTOR LOADS

A. Continuous Duty

Motor Load	Calculation
120V Exhaust fans	120V x 1.75A x 4 = 840.00VA
240V 40HP Fire pump	240V x 104A* x 1.732 = 43,230.72VA
240V 3HP Jockey pump	240V x 9.6A* x 1.732 = <u>3990.53VA</u>
	48,061.25VA

B. Non-Continuous Duty

Motor Load	Calculation
10HP motors-Int. dty/15mins.	240V x 28A* x 1.732 x 2 x .85** = 19,786.4VA

C. and D. NA

E. Total Motor Loads - [LINE LOAD - Add lines A. and B.] - [NEUTRAL LOAD - Total motor loads with neutral connections (120V)]

 * Table 430.248(1ϕ) and Table 430.250(3ϕ)
 ** Table 430.22(E)

TOTAL = 67,847.65VA

6.   <u>LINE LOAD</u>   <u>NEUTRAL LOAD</u>
                                                          (120V)
      67,847.65VA   840VA

7. MEDICAL EQUIPMENT - NA

8. INDUSTRIAL EQUIPMENT - NA

9. **HEATING** and **AIR-CONDITIONING (AC) EQUIPMENT** <NEC References - 220.50, 220.51, 220.60, 430.6(A)(1) and 440.6(A)>

ELECTRICAL HEATING UNITS

Heating Unit	Calculation
100kW (3ϕ)	100,000W
TOTAL HEAT =	100,000VA

AIR-CONDITIONING (AC) UNITS

AC Unit	Calculation
AC Rooftop units +	
Fan motors (1HP)	14,500VA x 5 + (240V x 4.2A* x 5 x 1.732) = 81,229.28VA
	TOTAL AC = 81,229.28VA

Line Load = 100,000VA + 20,784VA = 120,784VA
(Largest Load)   (Air Handlers)

Air Handlers (7.5HP and 10HP) - 240V x (22A* +28A*) x 1.732 = 20,784VA

*Table 430.250

9.	LINE LOAD	NETRAL LOAD
	120,784VA	0

10. LARGEST MOTOR

Fire Pump - 43,230.72VA x .25 (25 percent) = 10,807.68VA

10.	LINE LOAD	NEUTRAL LOAD
	10,807.68VA	0

TOTAL DEMAND LOAD (LINE and **NEUTRAL)** (List each computed line and neutral loads below and total lines 1. - 10.)

	LINE LOAD	NEUTRAL LOAD
1. General Lighting	45,000.00VA	36,000.0VA
2. Other Lighting Loads	34,225.50VA	14,550.0VA
3. Receptacle Loads	30,200.00VA	28,400.0VA
4. Kitchen Equipment	59,316.14VA	9768.2VA
5. Specific Loads	135,601.60VA	12,351.6VA
6. Motor Loads	67,847.65VA	840.0VA
7. Medical Equipment	--	--

8. Industrial Equipment	--	--
9. Heating and AC Equipment	120,784.00VA	0
10. Largest Motor	10,807.68VA	0
TOTAL =	503,782.57VA	101,909.8VA

OCCUPANCY'S OPERATING LINE VOLTAGE - 240V(3ϕ)
(Given operating voltage or as determined per test examination)

CALCULATE MINIMUM LINE and NEUTRAL LOADS
(Divide Total Demand Load [VA] by operating line voltage [V])

LINE LOAD = 503,782.57VA / 240V x 1.732 = 1211.95A

NEUTRAL LOAD = 101,909.8VA / 240V x 1.732 = ~~245.16A~~* 231.61A

*Where the neutral load exceeds 200A, NEC 220.61(B)(2) permits the load to be reduced by 70 percent. Complete the following to determine the Total Neutral Load.

(1) 245.16A – 200A = 45.16A x .70 = 31.61A

(2) 31.61A + 200A = 231.61A

Neutral Load = 231.61A

SIZE SERVICE (Size of service based on the calculated LINE LOAD)

SIZE SERVICE REQUIRED (minimum) 1600A
(Single rating or combination - Main overcurrent device(s) to total **1600A**)

SIZING FEEDER/SERVICE CONDUCTORS

Using NEC 310.10(H)(1) and Table 310.15(B)(16), as a minimum based on NEC 240.4(C), use three 1750 kcmil aluminum conductors (75°C) per phase which has an individual ampacity of 545A, as the service conductors.

FEEDER/SERVICE CONDUCTORS 3-1750 kcmil aluminum conductors per phase

SIZING NEUTRAL CONDUCTOR(S)

Based on the calculated neutral load and the use of parallel service conductors, the neutral conductors must also be installed in parallel per NEC 310.10(H)(1). Because the service conductors will consist of three parallel conductors per phase, the neutral conductors must also consist of three parallel conductors based on the calculated neutral load being divided by three which results to 77.2A (231.61A / 3). Per Table 310.15(B)(16), at 75°C, a 2 AWG aluminum conductor which has a rated ampacity of 90A will satisfy the calculated neutral load being distributed amongst three such conductors. However, because NEC 310.10(H)(1) only allows a

1/0 AWG conductor or larger to be installed in parallel, three 1/0 AWG aluminum conductors must be used instead as the parallel neutral conductors.

NEUTRAL CONDUCTOR(S) 3-1/0 AWG aluminum conductors

SIZING GROUNDING ELECTRODE CONDUCTOR

Based on the equivalent area of the selected parallel aluminum service conductors (5250 kcmil-[1750kcmil x 3]), per Table 250.66, a 250 kcmil aluminum grounding electrode conductor must be used.

Although NEC 250.24(C)(1) requires the neutral (grounded) conductor to be either the same size or larger than the grounding electrode conductors, the total circular mils of the three 1/0 AWG neutral conductors (316,800 cmils [105,600 cmils x 3]) exceeds the circular mils of a 250 kcmil conductor (250,000 cmils) per Table 8 of Chapter 9. As a result, this classifies the use of the three 1/0 AWG parallel neutral conductors as being larger than the 250 kcmil grounding electrode conductor.

GROUNDING ELECTRODE CONDUCTOR 1-250 kcmil aluminum

Warehouse

103. A three-story 288,000SF medical warehouse with a 2700SF business and operations office will be served by a 4W, 3ϕ, 208/120V wye (Y) connected electrical system. Calculate the service and neutral loads for this system based on the information provided. Over 70 percent of lighting load is nonlinear.

Business and Operations Office

53 - 2 x 4 fluorescent light fixtures (1.38A per ballast - 120V)
39 - Duplex receptacles (120V)
15 - Desktop Computers (1.7A - 120V/continuous)
20 - Laser Printers (3.2A - 120V)
 2 - Copier (9.3A - 208V-1ϕ)
 2 - Copiers (5.6A - 120V)
Beverage Machine (11.4A - 120V)
Vending/Snack Machine (8.3A - 120V)
Microwave - 1800W (120V)
Coffeemakers (2) - 956W (120V)
Water cooler - 840W (120V)

Warehouse

2 - Freight elevators - Intermittent duty/continuous - 60HP (208V-3ϕ)
1 - Freight elevators - Intermittent duty/30 min. - 60HP (208V-3ϕ)
3 - Fire Pumps - 60HP (208V-3ϕ)

3 - Jockey Pumps - 5HP (208V-3ϕ)
287 - Receptacles (120V)
300' - Plugmolds (continuous) (120V)
6 - Water coolers (840W-120V)
4 - Water Heaters (9.7kW-3ϕ)
Heating (225kW-3ϕ)
AC (188,000VA-3ϕ)
3 - Water Pumps (15HP-3ϕ)
36 - Exhaust fans (120V - 2.67A)
Outside Lighting (12,500VA/208V-1ϕ)

1. GENERAL LIGHTING or ACTUAL LIGHTING LOADS

General Lighting Load <NEC References - 220.12 and Table 220.12>

Business and Operations Office - 2700SF x 3.5VA	=	9450VA
Warehouse - 288,000SF x ¼VA	= ~~72,000VA~~	42,250VA
		51,700VA

APPLY DEMAND FACTORS FOR WAREHOUSES ONLY

a. First 12,500VA or less of above TOTAL (at 100%) = 12,500VA

b. 59,500 x .50 (at 50%) = 29,750VA
   (Remainder of TOTAL VA exceeding 12,500VA)*

   *(72,000VA (TOTAL) - 12,500 VA = 59,500VA)

   TOTAL (Lines a and b) = 42,250VA
   (Derated Lighting Loads)

   A. General Lighting Load - 9450VA x 1.25 + 42,250VA = 54,062.5VA

Actual Lighting Load

Type Fixture	VA rating	No. of Fixtures		TOTAL VA
Office				
Fluorescent light fixtures	120V x 1.38A x	53	=	8776.8

   B. Actual Lighting Load = ~~8776.8VA~~
      <Omit Actual Lighting Load - General Lighting Load Larger>

1.	LINE LOAD	NEUTRAL LOAD	
		Permitted Reduction	Prohibited Reduction
	54,062.5VA	--	51,700VA

2. OTHER LIGHTING LOADS

A. Sign/Outline (S/O) Lighting <NEC References - 220.12(F) and 600.5(A)>

Sign = 1200VA [Although not listed required per NEC 600.5(A)]
Total (Sign/Outline Lighting) = 1200VA (treat as neutral load)

B. Outside Lighting <NEC Reference - 220.18(B)>

Outside lighting - 12,500VA
Total (Outside Lighting) = 12,500VA (208V-1ϕ)

C. - E.   NA

F. Other Lighting (LINE and NEUTRAL) Loads Total [Add lines A. and B.]

Other Lighting Loads Total = 13,700VA
TOTAL = 13,700VA x 1.25 = 17,125VA

2.	LINE LOAD	NEUTRAL LOAD	
		(120V)	
		Permitted Reduction	Prohibited Reduction
	17,125VA	--	1200VA

3. RECEPTACLE LOADS

A(2) Non-continuous duty - General Purpose Receptacles (Warehouse) <NEC 220.14(I)>

(287 + 39*) x 180VA = 58,680VA

*Since the results of B. (1) is larger than B. (2) below, this method can be omitted thus allowing the receptacles [39] to be combined with the receptacles of step 3.A(2) [287] to obtain a smaller receptacle demand between the two non-continuous receptacle loads opposed to gathering a larger demand using two individual calculations.

APPLY DEMAND FACTORS
a. First 10,000VA (10kVA) (@100%) =   10,000VA
b. Remainder - 48,680VA x .50   =   24,340VA
                          Total =   34,340VA [Receptacle load per 3.A(2) & B.]

B. Non-continuous duty (Office) <NEC 220.14(K)>

(1) 39 x 180VA = 7020VA (Larger)

(2) 2700SF x 1VA/SF = 2700VA

C. Continuous duty <NEC References - 220.14(I), 215.2(A)(1) or 230.42(A)>

(Plugmolds) 300' x 180VA = 54,000VA x 1.25 = 67,500VA

D. Total Receptacle (NEUTRAL and LINE) Load

Total Receptacle loads - 34,340VA + 67,500VA = 101,840VA (LINE)
34,340VA + 54,000VA = 88,340VA (NEUTRAL)

3.	LINE LOAD		NEUTRAL LOAD	
			Permitted Reduction	Prohibited Reduction
	101,840VA		88,340VA	--

4. KITCHEN EQUIPMENT - NA

5. SPECIFIC LOADS

Type Load	Calculation		
120V			
Beverage Machine	120V x 11.4A	=	1368.0VA
Coffeemakers	956VA x 2	=	1912.0VA
Copiers*	120V x 5.6A x 2	=	1344.0VA
Desktop computers*	120V x 1.7A x 15 x 1.25	=	3825.0VA
Laser Printers*	120V x 3.2A x 20	=	7680.0VA
Microwave		=	1800.0VA
Vending/Snack Machine	120V x 8.3A	=	996.0VA
Water coolers	840VA x 7	=	5880.0VA
			24,805.0VA
208V-1ϕ			
Copiers	208V x 9.3A x 2	=	3868.8VA
208V-3ϕ			
Water Heaters	9700VA x 4	=	38,800.0VA
		TOTAL =	67,473.8VA

5.	LINE LOAD		NEUTRAL LOAD (120V)	
			Permitted Reduction	Prohibited* Reduction
	67,473.8VA		11,956VA	12,849VA

6. MOTOR LOADS

A. Continuous Duty

Motor Load	Calculation		
Exhaust fans (120V)	120V x 2.67A x 36	=	11,534.40VA
Fire pumps (60HP)	208V x 169A* x 1.732 x 3	=	182,649.79VA
Jockey pumps (5HP)	208V x 16.7A* x 1.732 x 3	=	18,048.83VA
Water pumps (15HP)	208V x 46.2A* x 1.732 x 3	=	49,931.48VA
			262,164.50VA

B. and C. NA

D. Elevators - Intermittent Duty Cycle (3)

60HP motors-Intermit. dty/cont. - 208V x 169A* x 1.732 x 2 x 1.40** = 170,473.14VA
60HP motors-Intermit. dty/30min. - 208V x 169A* x 1.732 x .90** = 54,794.94VA
~~225,268.08VA~~

225,268.08VA x .90** = 202,741.27VA (Demand Load)
**Apply Demand Factor (Table 620.14) for Elevators (3 @ .90) (assumed - not under constant load)

E. Total Motor Loads - [LINE LOAD - Add lines A. and D.] - [NEUTRAL LOAD - Total motor loads with neutral connections (120V)]

*Table 430.250 **Table 430.22(E)   TOTAL =   464,905.77VA

6.   LINE LOAD

	NEUTRAL LOAD	
	Permitted Reduction	Prohibited Reduction
464,905.77VA	11,534.4VA	--

7. MEDICAL EQUIPMENT - NA

8. INDUSTRIAL EQUIPMENT - NA

9. HEATING and AIR-CONDITIONING (AC) EQUIPMENT <NEC References - 220.50, 220.51, 220.60, 430.6(A)(1) and 440.6(A)>

HEAT = 225kW (225,000VA) AC = 188,000VA
Line Load = 225,000VA (Largest)

Line Load = 225,000VA (Largest)

9.   LINE LOAD

	NEUTRAL LOAD	
	Permitted Reduction	Prohibited Reduction
225,000VA	0	--

10. LARGEST MOTOR

One 60HP Motor (highest current per NEC 430.17). Although there are six 60HP motors with similar ratings only one is considered.

208V x 169A* x 1.732 x .25 = 15,220.82VA
*Table 430.250

10.	LINE LOAD	NEUTRAL LOAD	
		Permitted Reduction	Prohibited Reduction
	15,220.82VA	--	--

TOTAL DEMAND LOAD (LINE and NEUTRAL) (List each computed line and neutral loads below and total lines 1. – 10.)

		LINE LOAD	NEUTRAL LOAD	
			Permitted Reduction	Prohibited Reduction
1.	General Lighting	54,062.50VA	--	51,700VA
2.	Other Lighting Loads	17,125.00VA	--	1200VA
3.	Receptacle Loads	101,840.00VA	88,340.0VA	--
4.	Kitchen Equipment	--	--	--
5.	Specific Loads	67,473.80VA	11,956.0VA	12,849VA
6.	Motor Loads	464,905.77VA	11,534.4VA	--
7.	Medical Equipment	--	--	--
8.	Industrial Equipment	--	--	--
9.	Heating and AC Equip.	225,000.00VA	0	--
10.	Largest Motor	15,220.82VA	0	--
	TOTAL =	945,627.89VA	111,830.4VA	65,749VA

OCCUPANCY'S OPERATING LINE VOLTAGE - <u>208V</u>(3φ)
(Given operating voltage or as determined per test examination)

CALCULATE MINIMUM LINE AND NEUTRAL LOADS
(Divide Total Demand Load [VA] by operating line voltage [V])

<u>LINE LOAD</u> = 945,627.89VA / 208V x 1.732 = <u>2624.88A</u>
<u>NEUTRAL LOAD</u>
        Permitted = 111,830.4VA / 208V x 1.732 = ~~310.42A~~* <u>277.29A</u>
        Prohibited = 65,749VA / 208V x 1.732 = <u>182.51A</u>

*Where the (permitted) neutral load exceeds 200A, NEC 220.61(B)(2) permits the load to be reduced by 70 percent. Complete the following to determine the Total Neutral Load.

(1) 310.42A - 200A = 110.42A x .70 = 77.29A   (2) 77.29A + 200A = 277.29A

TOTAL NEUTRAL LOAD = 277.29A + 182.51A = <u>459.80A</u>

PART IV - Optional Feeder and Service Load Calculations

The following procedures for calculating service loads for residential and commercial buildings are based on applicable sections of Part IV of Article 220 as referenced.

220.82(B)(1) and **(2)** - General Loads [(One-Family) Dwelling Unit]

Refer to **(Worksheet B** - Volume 4) OPTIONAL LOAD CALCULATIONS FOR ONE-FAMILY DWELLING for related questions.

104. Re-calculate the demand load for the single-family dwelling in question No. 81. using the optional load calculation.

1. General Lighting and Receptacle Loads <NEC 220.82(B)(1)> – (Open porches, garages, unused or unfinished spaces not adaptable for future use not included)

Because the unfinished basement is adaptable for future use it must be included with the total square footage of the house.

$$2435SF + 800SF \text{ x } 3VA = 9705VA$$

2. Small-Appliance (Portable) **and Laundry Circuit Load** <NEC 220.82(B)(2)>

$$1500VA \text{ x } 6 = 9000VA$$

3. Appliances and Motor Loads <(NEC 220.82(B)(3) and (4)> - NA

4. Apply Demand Formula <(NEC 220.82(B)>

TOTAL = <u>18,705</u>VA
[Loads **1.** – **3.** added]

a. First 10kVA (10,000VA) or less of above TOTAL		= 10,000VA
b. 8705VA	x 40	= <u>3482</u>VA
c. Total		= 13,482VA

Using the optional load calculation, the general lighting, receptacle, small-appliance and laundry demand load for this house is 13,482VA.

220.82(C)(1) - (6) - Heating and Air Conditioning Load [(One-Family) Dwelling Unit]
(105. - 110.) [6]

105. A heat pump unit without the use of supplemental electric heating is being used as the sole means for heating and cooling a one-family home. If the heat pump is rated for 7200VA determine the load of the unit applying the provisions of NEC 220.82(C)(2).

According to NEC 220.82(C)(2), 100 percent of the nameplate rating of a heat pump must be applied when the heat pump will be utilized without the use of any supplemental heating. Therefore, the load of this unit is 7200VA.

106. If the same heat pump unit in question No. 105. was supplemented with a 10kW electric heating unit determine the load if both units were capable of operating at the same time.

According to NEC 220.82(C)(3), when a heat pump and supplemental electric heating will operate at the same time, the load must be determined based on 100 percent of the heat pump and 65 percent of the electric heating. As a result, the overall load amounts to,

$$7200VA + (10,000VA \times .65) = 13,700VA$$

107. Per NEC 220.82(C)(4), if three wall-mounted heating units rated for 4500W, 5000W, and 6200W with separate thermostats were used to heat a dwelling unit, what would the heating load result to?

According to NEC 220.82(C)(4), if less than four separately controlled electric space heating units are utilized, a 65 percent demand can be applied towards the overall heating load. Therefore,

$$(4500VA + 5000VA + 6200VA) \times .65 = 10,205VA$$

108. Re-calculate the heating load in question No. 107. if two separately controlled electric space heating units rated for 7500W were added to the heating load.

According to NEC 220.82(C)(5), when more than four separately controlled electric space heating units are utilized, a 40 percent demand can be applied towards the overall heating load. Therefore,

$$(4500VA + 5000VA + 6200VA + 7500VA + 7500VA) \times .40 = 12,280VA$$

109. Two electric thermal storage systems are being used to heat a home. If both systems are rated to supply a continuous heating load of 24,500W determine the heating load for this home.

According to NEC 220.82(C)(6), 100 percent of the nameplate rating(s) of an electric thermal storage system must be applied. Therefore, the heating for this home would result to 24,500VA.

110. If a 21,350VA cooling load was considered with question No. 108., what would the heating and cooling demand result to?

Based on the provisions of NEC 220.82(C) which states "the largest of the following six selections [(1) – (6)] shall be included:" and the fact that both cooling and heating loads will never operate at the same time, the cooling load (21,350VA) compared to the heating load (12,280VA) would determine the heating and cooling demand.

220.82(B) - General Loads and **220.82(C) - Heating and Air Conditioning Load (One-Family) Dwelling Unit) (111. - 114.) [4]**

Refer to **(Worksheet B** - Volume 4) OPTIONAL LOAD CALCULATIONS FOR ONE-FAMILY DWELLING for related questions.

111. Determine the optional load for a one-family dwelling served by a single-phase, 3-wire, 120/240V service, with a general load calculation of 43.5kVA and an electric heating load of 23.7kVA?

Based on the provisions of NEC 220.82(B) the general load calculation (43.5kVA) must be reduced per demand factor as follows since this load represents the dwelling at 100 percent:

Apply Demand Formula <NEC 220.82(B)> to 43.5kVA(43,500VA) General Loads

a. First 10kVA (10,000VA) or less of above TOTAL   =   10,000VA
b. 33,500VA (Remainder of TOTAL)   x .40   =   13,400VA
c. Total   =   23,400VA

Heating and Air-Conditioning Load <NEC 220.82(C)> 23.7kVA(23,700VA) Heating Load

(1) NA
(2) NA
(3) NA
(4) Heat = 23,700VA x 65% (.65) = 15,405VA
(5) NA
(6) NA

TOTAL DEMAND LOAD

		LINE LOAD
General Loads	=	23,400VA
Heating and Air-Conditioning (AC) Equipment	=	15,405VA
TOTAL DEMAND LOAD	=	38,805VA

DWELLING'S OPERATING LINE VOLTAGE - 240V
(Given operating voltage or as determined per test examination)

CALCULATE MINIMUM LINE LOAD
(Divide Total Demand Load **[VA]** by operating line voltage **[V]**)

LINE LOAD = 38,805VA / 240V = **161.69A**

112. Re-calculate the service load and determine the size service needed for the single-family dwelling in question No. 82. using the optional load calculation.

NEC 220.82(B) - General Loads (1. – 4.)

1. General Lighting and Receptacle Loads <NEC 220.82(B)(1)> - (Open porches, garages, unused or unfinished spaces not adaptable for future use not included)

$$4200SF \times 3VA = 12,600VA$$

2. Small-Appliance (Portable) **and Laundry Circuit Load** <NEC 220.82(B)(2)>

$$1500VA \times 7 = 10,500VA$$

3. Appliances and Motor Loads <NEC 220.82(B)(3) and (4)>

Appliances and Motors Loads	VA rating
Cooktop	10,000
Dishwasher	1000
Disposal	960
Dryer	4500
Microwave oven	1600
Wall-mounted ovens - (8.5kW - 2)	17,000
Water Heater	5000
Blower motors - (745VA - 2)	1490
HVL (1500W - 4)	6000
Total Appliances VA rating =	47,550

4. Apply Demand Formula <NEC 220.82(B)>

$$TOTAL = \underline{70,650VA}$$
[Loads **1. - 3.** added]

   a. First 10kVA (10,000VA) or less of above TOTAL   = 10,000VA
   b. 60,650VA (Remainder of TOTAL) x .40   = 24,260VA
   c. Total   = 34,260VA

5. Heating and Air-Conditioning Load <NEC 220.82(C)>

   (1) AC = 230V x 29A + 230V x 1.5A = 7015VA
             230V x 23.5A + 230V x 1.5A = 5750VA
                                       12,756VA

   (2) and (3) NA

(4) Heat = 15kW (15,000VA) x 2 = 30kW (30,000VA)

30kW(30,000VA) x 65% (.65) = 19,500VA

(5) and (6) NA

TOTAL DEMAND LOAD
(Add 4c. and 5. [the largest selection])

LINE LOAD

4c. General Lighting and Receptacle, Small-
Appliances and Laundry Circuit Loads
and Appliances   = 34,260VA

5. Heating and Air-Conditioning (AC) Equipment   = 19,500VA

TOTAL DEMAND LOAD   = 53,760VA

DWELLING'S OPERATING LINE VOLTAGE - 230V
(Given operating voltage or as determined per test examination)

CALCULATE MINIMUM LINE LOAD
(Divide Total Demand Load **[VA]** by operating line voltage **[V]**)

LINE LOAD = 53,760VA / 230V = **233.74A**

The service load for this single-family dwelling using the optional calculation is as calculated per line load.
COMPARISON OF RESULTS

	Standard Calculation Question No. 82	**Optional Calculation** Question No. 112
LINE LOAD/SIZE SERVICE	317.88A/350A	233.74A/250A

113. Re-calculate the service load and determine the size service needed for the single-family dwelling in question No. 83. using the optional load calculation.

NEC 220.82(B) - General Loads (1. – 4.)

1. General Lighting and Receptacle Loads <NEC 220.82(B)(1)> – (Open porches, garages, unused or unfinished spaces not adaptable for future use not included)

9700SF x 3VA = 29,100VA

2. Small-Appliance (Portable) **and Laundry Circuit Load** <NEC 220.82(B)(2)>

1500VA x 10 = 15,000VA

3. Appliances and Motors Loads <NEC 220.82(B)(3) and (4)>

Appliances and Motors Loads	VA rating
Cooktop	6500
Cooktop	6800
Dishwashers - (1kW - 2)	2000
Disposal	960
Disposal	780
Dryers - (6.7kW - 2)	13,400
Microwave ovens - (1600W - 2)	3200
Microwave oven	1250
Wall-mounted ovens - (5.5kW - 2)	11,000
Wall-mounted oven	6400
Water Heaters - (4.5kW - 3)	13,500
Blower mtrs. - (2) (240V x 14.8A x 2)	7104
Blower mtrs. - (2) (240V x 10.3A x 2)	4944
Blower mtr. - (240V x 10.3A)	2472
Circ. mtr. - (Pool) (240V x 10A*)	2400
Circ. mtr. - (Spa) (120V x 20A*)	2400
Cleaning mtr. - (Pool) (240V x 6.9A*)	1656
Deep fryer	2200
Freezer - (120V x 13.3A)	1596
Fountain Pump - (120V x 20A*)	2400
Gate Opener - (120V x 9.8A*)	1176
HVL - (1500W - 6)	9000
Heater (Spa)	5500
Rotisserie	2000
Sump Pump (120V x 13.8A*)	1656
Trash compactors (756VA - 2)	1512
Vacuum mtr. (120V x 16A*)	1920

*Table 430.248

Total Appliances VA rating = 115,726

4. Apply Demand Formula <NEC 220.82(B)>

$$TOTAL = \underline{159,826}VA$$

[Loads **1.** – **3.** added]

a. First 10kVA (10,000VA) or less of above TOTAL = 10,000.0VA
b. 159,826VA (Remainder of TOTAL)   x .40   = 63,930.4VA
c. Total   = 73,930.4VA

5. Heating and Air-Conditioning Load <NEC 220.82(C)>

(1) AC Load (5 units)

      Compressors   **Fan Motors**

     (240V x 25.3A + 240V x 1.2A) x 2 = 12,720VA

     (240V x 20.2A + 240V x 1.2A) x 2 = 10,272VA

     240V x 14.4A + 240V x 1.1A =   <u>3720VA</u>

                                         26,712VA

(2) - (4) NA

(5) Heating Load (5 units)

     20kW (20,000VA) x 2 = 40,000VA

     15kW (15,000VA) x 2 = 30,000VA

     10kW (10,000VA)   = <u>10,000VA</u>

                                   80,000VA

     80kW (80,000VA) x 40% (.40) = 32,000VA

(6) NA

TOTAL DEMAND LOAD

(Add 4c. and 5. [the largest selection])

                                                             LINE LOAD

4c. General Lighting and Receptacle, Small-
Appliances and Laundry Circuit Loads
and Appliances   =   73,930.4VA

5. Heating and Air-Conditioning (AC) Equipment =   <u>32,000.0VA</u>

         TOTAL DEMAND LOAD   =   105,930.4VA

DWELLING'S OPERATING LINE VOLTAGE - <u>240V</u>

(Given operating voltage or as determined per test examination)

CALCULATE MINIMUM LINE and NEUTRAL LOADS

(Divide Total Demand Load [VA] by operating line voltage [V])

       LINE LOAD = <u>105,930.4VA</u> / 240V = **441.38A**

The service load for this single-family dwelling using the optional calculation is as calculated per line load.

COMPARISON OF RESULTS

	Standard Calculation Question No. 83.	**Optional Calculation** Question No. 113.
LINE LOAD	753.77A	441.38A
SIZE SERVICE	800A	450A

114. Recalculate the service load and determine the size service needed for the single-family dwelling in question No. 87. using the optional load calculation. Assume all dryers at 5000W.

650SF unit - 208/120V 1-phase

NEC 220.82(B) – General Loads (1. – 4.)

1. General Lighting and Receptacle Loads <NEC 220.82(B)(1)> - (Open porches, garages, unused or unfinished spaces not adaptable for future use not included)

$$650SF \times 3VA = 1950VA$$

2. Small-Appliance and Laundry Circuit[1] Load <NEC 220.82(B)(2)>

$$1500VA \times 2 = 3000VA$$
[1]650SF units (15) not equipped with washer/dryer connections.

3. Appliances and Motors Loads <NEC 220.82(B)(3) and (4)>

Appliances and Motors Loads	VA rating
Dishwasher	900
Disposal	600
Heat-Vent-Light	1300
Microwave	1000
Range	6100
Water Heater	3380
Blower	1044
Total Appliances VA rating =	14,324

4. Apply Demand Formula <NEC 220.82(B)>

$$TOTAL = \underline{19,274VA}$$
[Loads **1.** – **3.** added]

   a. First 10kVA (10,000VA) or less of above TOTAL = 10,000.0VA
   b. 9274VA (Remainder of TOTAL)   x .40   =   3709.6VA
   c. Total   = 13,709.6VA

5. Heating and Air-Conditioning Load <NEC 220.82(C)>

   (1) AC = 3648VA
   (2) and (3) NA
   (4) Heat = 10kW(10,000VA)
          10kW(10,000VA) x 65% (.65) = 6500VA
   (5) and (6) NA

TOTAL DEMAND LOAD
(Add 4c. and 5. [the largest selection])

                                                                    LINE LOAD

4c. General Lighting and Receptacle, Small-
    Appliances and Laundry Circuit Loads
    and Appliances   =   13,709.6VA

5. Heating and Air-Conditioning (AC) Equipment =   6500.0VA

              TOTAL DEMAND LOAD   =   20,209.6VA

DWELLING'S OPERATING LINE VOLTAGE - 208V
(Given operating voltage or as determined per test examination)

CALCULATE MINIMUM LINE LOAD
(Divide Total Demand Load **[VA]** by operating line voltage **[V]**)

LINE LOAD = 20,209.6VA / 208V = **97.16A**

The service load for this single-family dwelling using the optional calculation is as calculated per line load.

COMPARISON OF RESULTS

	Standard Calculation	**Optional Calculation**
	Question No. 87.	Question No. 114.
LINE LOAD	121.41A	97.16A
SIZE SERVICE	125A	100A

650SF unit - 240/120V 1-phase

NEC 220.82(B) – General Loads (1. – 4.)

1. General Lighting and Receptacle Loads <NEC 220.82(B)(1)> - (Open porches, garages, unused or unfinished spaces not adaptable for future use not included)

650SF x 3VA = 1950VA

2. Small-Appliance and Laundry Circuit[1] Load <NEC 220.82(B)(2)>

1500VA x 2 = 3000VA

[1]650SF units (15) not equipped with washer/dryer connections.

3. Appliances and Motors Loads <NEC 220.82(B)(3) and (4)>

Appliances and Motors Loads	VA rating
Dishwasher	900
Disposal	600
Heat-Vent-Light	1300
Microwave	1000
Range	8100
Water Heater	4500
Blower	785
Total Appliances VA rating =	17,185

4. Apply Demand Formula <NEC 220.82(B)>

$$\text{TOTAL} = \underline{22,135}\text{VA}$$
[Add loads **1. – 3.**]

a. First 10kVA (10,000VA) or less of above TOTAL = 10,000.0VA
b. 12,135VA (Remainder of TOTAL) x .40 = 4854.0VA
c. Total = 14,854.0VA

5. Heating and Air-Conditioning Load <NEC 220.82(C)>

(1) AC = 2743VA
(2) and (3) NA
(4) Heat = 13.3kW(13,300VA) - 13.3kW(13,300VA) x 65% (.65) = 8645VA
(5) and (6) NA

TOTAL DEMAND LOAD
(Add 4c. and 5. [the largest selection])

LINE LOAD

4c. General Lighting and Receptacle, Small-
Appliances and Laundry Circuit Loads
and Appliances = 14,854.0VA

5. Heating and Air-Conditioning (AC) Equipment = 8645.0VA

TOTAL DEMAND LOAD = 23,499.0VA

DWELLING'S OPERATING LINE VOLTAGE - 240V
(Given operating voltage or as determined per test examination)

CALCULATE MINIMUM LINE LOAD
(Divide Total Demand Load [VA] by operating line voltage [V])

LINE LOAD = <u>23,499</u>VA / <u>240</u>V = **97.91A**

The service load for this single-family dwelling using the optional calculation is as calculated per line load.

<u>COMPARISON OF RESULTS</u>

	Standard Calculation Question No. 87.	**Optional Calculation** Question No. 114.
LINE LOAD	127.59A	97.91A
SIZE SERVICE	150A	100A

900SF unit - 208/120V 1-phase

NEC 220.82(B) – General Loads (1. – 4.)

1. General Lighting and Receptacle Loads <NEC 220.82(B)(1)> - (Open porches, garages, unused or unfinished spaces not adaptable for future use not included)

$$900SF \times 3VA = 2700VA$$

2. Small-Appliance* and Laundry Circuit** Load <NEC 220.82(B)(2)>

$$1500VA \times (2^* + 1^{**}) = 4500VA$$

3. Appliances and Motors Loads <NEC 220.82(B)(3) and (4)>

Appliances and Motors Loads	VA rating
Dishwasher	900
Disposal	600
Dryer	5000
Heat-Vent-Light	1600
Microwave	1200
Range	7800
Water Heater	3380
Blower	1176

Total Appliances VA rating = 21,656

4. Apply Demand Formula <NEC 220.82(B)>

$$TOTAL = \underline{28,856}VA$$
[Loads **1.** – **3.** added]

   a. First 10kVA (10,000VA) or less of above TOTAL = 10,000.0VA
   b. 18,856VA (Remainder of TOTAL)   x .40   =   <u>7542.4VA</u>
   c. Total   = 17,542.4VA

5. Heating and Air-Conditioning Load <NEC 220.82(C)>

   (1) AC = 4560VA
   (2) and (3) NA
   (4) Heat = 12kW(12,000VA) - 12kW(12,000VA) x 65% (.65) = 7800VA
   (5) and (6) NA

TOTAL DEMAND LOAD
(Add 4c. and 5. [the largest selection])

LINE LOAD

4c. General Lighting and Receptacle, Small-
Appliances and Laundry Circuit Loads
and Appliances   = 17,542.4VA

5. Heating and Air-Conditioning (AC) Equipment   =   7800.0VA

TOTAL DEMAND LOAD   = 25,342.4VA

DWELLING'S OPERATING LINE VOLTAGE - 208V
(Given operating voltage or as determined per test examination)

CALCULATE MINIMUM LINE LOAD
(Divide Total Demand Load **[VA]** by operating line voltage **[V]**)

LINE LOAD = 25,342.4VA / 208V = **121.84A**

The service load for this single-family dwelling using the optional calculation is as calculated per line load.

COMPARISON OF RESULTS

	Standard Calculation Question No. 87	**Optional Calculation** Question No. 114.
LINE LOAD	167.98A	121.84A
SIZE SERVICE	175A	125A

900SF unit - 240/120V 1-phase

NEC 220.82(B) – General Loads (1. – 4.)

1. General Lighting and Receptacle Loads <NEC 220.82(B)(1)> - (Open porches, garages, unused or unfinished spaces not adaptable for future use not included)

900SF x 3VA = 2700VA

2. Small-Appliance* and Laundry Circuit** Load <NEC 220.82(B)(2)>

1500VA x (2* + 1**) = 4500VA

3. Appliances and Motors Loads <NEC 220.82(B)(3) and (4)>

Appliances and Motors Loads	VA rating
Dishwasher	900
Disposal	600
Dryer	5000
Heat-Vent-Light	1600
Microwave	1200
Range	10,400
Water Heater	4500
Blower	884

Total Appliances VA rating = 25,084

4. Apply Demand Formula <NEC 220.82(B)>

TOTAL = 32,284VA
[Loads **1.** – **3.** added]

a. First 10kVA (10,000VA) or less of above TOTAL = 10,000.0VA
b. 22,284VA (Remainder of TOTAL)   x .40   =   8913.6VA
c. Total   = 18,913.6VA

5. Heating and Air-Conditioning Load <NEC 220.82(C)>

(1) AC = 3429VA
(2) and (3) NA
(4) Heat = 16kW (16,000VA) - 16kW (16,000VA) x 65% (.65) = 10,400VA
(5) and (6) NA

TOTAL DEMAND LOAD
(Add 4c. and 5. [the largest selection])

LINE LOAD

4c. General Lighting and Receptacle, Small-
Appliances and Laundry Circuit Loads
and Appliances   = 18,913.6VA

5. Heating and Air-Conditioning (AC) Equipment   = 10,400.0VA

TOTAL DEMAND LOAD   = 29,313.6VA

DWELLING'S OPERATING LINE VOLTAGE - 240V
(Given operating voltage or as determined per test examination)

CALCULATE MINIMUM LINE LOAD
(Divide Total Demand Load [**VA**] by operating line voltage [**V**])

$$\text{LINE LOAD} = \underline{29,313.6}\text{VA} / \underline{240}\text{V} = \mathbf{122.14A}$$

The service load for this single-family dwelling using the optional calculation is as calculated per line load.

COMPARISON OF RESULTS

	Standard Calculation Question No. 87.	**Optional Calculation** Question No. 114.
LINE LOAD	171.27A	122.14A
SIZE SERVICE	175A	125A

1200SF unit - 208/120V 1-phase

NEC 220.82(B) – General Loads (1. – 4.)

1. **General Lighting and Receptacle Loads** <NEC 220.82(B)(1)> – (Open porches, garages, unused or unfinished spaces not adaptable for future use not included)

$$1200\text{SF} \times 3\text{VA} = 3600\text{VA}$$

2. **Small-Appliance* and Laundry Circuit** Load** <NEC 220.82(B)(2)>

$$1500\text{VA} \times (2^* + 1^{**}) = 4500\text{VA}$$

3. **Appliances and Motors Loads** <NEC 220.82(B)(3) and (4)>

Appliances and Motors Loads	VA rating
Cooktop	6300
Dishwasher	1250
Disposal	900
Dryer	5000
Heat-Vent-Light	1600
Microwave	1600
Trash Compactor	780
Wall-mounted oven	5400
Water Heater	3380
Blower	1320

Total Appliances VA rating = 27,530

4. Apply Demand Formula <NEC 220.82(B)>

$$TOTAL = \underline{35,630}VA$$
[Loads **1.** – **3.** added]

a. First 10kVA (10,000VA) or less of above TOTAL = 10,000VA
b. 25,630VA (Remainder of TOTAL)   x 40   = 10,252VA
c. Total   = 20,252VA

5. Heating and Air-Conditioning Load <(NEC 220.82(C)>

(1) AC = 5472VA
(2) and (3) NA
(4) Heat = 15kW (15,000VA) - 15kW (15,000VA) x 65% (.65) = 9750VA
(5) and (6) NA

TOTAL DEMAND LOAD
(Add 4c. and 5. [the largest selection])

LINE LOAD

4c. General Lighting and Receptacle, Small-
     Appliances and Laundry Circuit Loads
     and Appliances   = 20,252VA

5. Heating and Air-Conditioning (AC) Equipment   = 9750VA

         TOTAL DEMAND LOAD   = 30,002VA

DWELLING'S OPERATING LINE VOLTAGE - 208V
(Given operating voltage or as determined per test examination)

CALCULATE MINIMUM LINE LOAD
(Divide Total Demand Load [VA] by operating line voltage [V])

LINE LOAD = 30,002VA / 208V = **144.24A**

The service load for this single-family dwelling using the optional calculation is as calculated per line load.

COMPARISON OF RESULTS

	Standard Calculation Question No. 87	**Optional Calculation** Question No. 114.
LINE LOAD	197.44A	144.24A
SIZE SERVICE	200A	150A

1200SF unit - 240/120V 1-phase

NEC 220.82(B) – General Loads (1. – 4.)

1. **General Lighting and Receptacle Loads** <NEC 220.82(B)(1)> - (Open porches, garages, unused or unfinished spaces not adaptable for future use not included)

$$1200SF \times 3VA = 3600VA$$

2. **Small-Appliance* and Laundry Circuit** Load** <NEC 220.82(B)(2)>

$$1500VA \times (2^* + 1^{**}) = 4500VA$$

3. **Appliances and Motors Loads** <NEC 220.82(B)(3) and (4)>

Appliances and Motors Loads	VA rating
Cooktop	8400
Dishwasher	1250
Disposal	900
Dryer	5000
Heat-Vent-Light	1600
Microwave	1600
Trash Compactor	780
Wall-mounted oven	7180
Water Heater	4500
Blower	992

Total Appliances VA rating = 32,202

4. **Apply Demand Formula** <NEC 220.82(B)>

$$TOTAL = \underline{40,302VA}$$
[Loads **1.** – **3.** added]

   a. First 10kVA (10,000VA) or less of above TOTAL = 10,000.0VA
   b. 30,302VA (Remainder of TOTAL)   x .40   = 12,120.8VA
   c. Total   = 22,120.8VA

5. **Heating and Air-Conditioning Load** <(NEC 220.82(C)>

   (1) AC = 4114VA
   (2) and (3) NA
   (4) Heat = 20kW (20,000VA) - 20kW (20,000VA) x 65% (.65) = 13,000VA
   (5) and (6) NA

TOTAL DEMAND LOAD
(Add 4c. and 5. [the largest selection])

LINE LOAD

4c. General Lighting and Receptacle, Small-
   Appliances and Laundry Circuit Loads
   and Appliances   = 22,120.8VA

5. Heating and Air-Conditioning (AC) Equipment   = 13,000.0VA

                    TOTAL DEMAND LOAD   = 35,120.8VA

DWELLING'S OPERATING LINE VOLTAGE - 240V
(Given operating voltage or as determined per test examination)

CALCULATE MINIMUM LINE LOAD
(Divide Total Demand Load **[VA]** by operating line voltage **[V]**)

LINE LOAD = 35,120.8VA VA / 240V = **146.34A**

The service load for this single-family dwelling using the optional calculation is as calculated per line load.

COMPARISON OF RESULTS

	Standard Calculation Question No. 87.	Optional Calculation Question No. 114.
LINE LOAD	204.6A	146.34A
SIZE SERVICE	225A	150A

1400SF unit - 208/120V 1-phase

NEC 220.82(B) – General Loads (1. – 4.)

1. **General Lighting and Receptacle Loads** <NEC 220.82(B)(1)> - (Open porches, garages, unused or unfinished spaces not adaptable for future use not included)

   1400SF x 3VA = 4200VA

2. **Small-Appliance* and Laundry Circuit** Load** <NEC 220.82(B)(2)>

   1500VA x (2* + 1**) = 4500VA

3. **Appliances and Motors Loads** <NEC 220.82(B)(3) and (4)>

Appliances and Motors Loads	VA rating
Cooktop	6700
Dishwasher	1250
Disposal	900
Dryer	5000
Heat-Vent-Light	1750
Microwave	1800
Trash Compactor	780
Wall-mounted oven	5800
Water Heater	3380
Blower	<u>1416</u>

Total Appliances VA rating = 28,776

4. **Apply Demand Formula** <NEC 220.82(B)>

TOTAL = <u>37,476</u>VA
[Loads **1.** – **3.** added]

a. First 10kVA (10,000VA) or less of above TOTAL = 10,000.0VA
b. 27,476VA (Remainder of TOTAL)   x .40   = <u>10,990.4VA</u>
c. Total   = 20,990.4VA

5. **Heating and Air-Conditioning Load** <NEC 220.82(C)>

(1) AC = 6384VA
(2) and (3) NA
(4) Heat = 18kW (18,000VA) - 18kW (18,000VA) x 65% (.65) = 11,700VA
(5) and (6) NA

TOTAL DEMAND LOAD
(Add 4c. and 5. [the largest selection])

LINE LOAD

4c. General Lighting and Receptacle, Small-
        Appliances and Laundry Circuit Loads
        and Appliances   = 20,990.4VA

5. Heating and Air-Conditioning (AC) Equipment   = <u>11,700.0VA</u>

            TOTAL DEMAND LOAD   = 32,690.4VA

DWELLING'S OPERATING LINE VOLTAGE - <u>208</u>V
(Given operating voltage or as determined per test examination)

CALCULATE MINIMUM LINE LOAD
(Divide Total Demand Load **[VA]** by operating line voltage **[V]**)

$$\text{LINE LOAD} = \underline{32,690.4}\text{VA} / \underline{208}\text{V} = \mathbf{\underline{157.17}\text{A}}$$

The service load for this single-family dwelling using the optional calculation is as calculated per line load.

COMPARISON OF RESULTS

	Standard Calculation Question No. 87.	Optional Calculation Question No. 114.
LINE LOAD/ SIZE SERVICE	217.1A/225A	157.17A/175A

<u>1400SF unit - 240/120V 1-phase</u>

NEC 220.82(B) – General Loads (1. – 4.)

1. **General Lighting and Receptacle Loads** <NEC 220.82(B)(1)> - (Open porches, garages, unused or unfinished spaces not adaptable for future use not included)

$$1400\text{SF x } 3\text{VA} = 4200\text{VA}$$

2. **Small-Appliance* and Laundry Circuit** Load** <NEC 220.82(B)(2)>

$$1500\text{VA x } (2^* + 1^{**}) = 4500\text{VA}$$

3. **Appliances and Motors Loads** <NEC 220.82(B)(3) and (4)>

Appliances and Motors Loads	VA rating
Cooktop	8900
Dishwasher	1250
Disposal	900
Dryer	5000
Heat-Vent-Light	1750
Microwave	1800
Trash Compactor	780
Wall-mounted oven	7700
Water Heater	4500
Blower	<u>1065</u>

Total Appliances VA rating   = 33,645

4. Apply Demand Formula <NEC 220.82(B)>

$$\text{TOTAL} = \underline{42,345\text{VA}}$$
[Loads **1.** – **3.** added]

  a. First 10kVA (10,000VA) or less of above TOTAL = 10,000.0VA
  b. 32,345VA (Remainder of TOTAL)   x .40   = <u>12,938.0VA</u>
  c. Total   = 22,938.0VA

5. Heating and Air-Conditioning Load <NEC 220.82(C)>

  (1) AC = 4800VA
  (2) and (3) NA
  (4) Heat = 24kW (24,000VA) - 24kW (24,000VA) x 65% (.65) = 15,600VA
  (5) and (6) NA

TOTAL DEMAND LOAD
(Add 4c. and 5. [the largest selection])

LINE LOAD

4c. General Lighting and Receptacle, Small-
    Appliances and Laundry Circuit Loads
    and Appliances   = 22,938.0VA

5. Heating and Air-Conditioning (AC) Equipment   = <u>15,600.0VA</u>

          TOTAL DEMAND LOAD   = 38,538.0VA

DWELLING'S OPERATING LINE VOLTAGE - <u>240V</u>
(Given operating voltage or as determined per test examination)

CALCULATE MINIMUM LINE LOAD
(Divide Total Demand Load [**VA**] by operating line voltage [**V**])

LINE LOAD = <u>38,538.0</u>VA / <u>240</u>V = **160.58A**

The service load for this single-family dwelling using the optional calculation is as calculated per line load.

COMPARISON OF RESULTS

	Standard Calculation Question No. 87.	**Optional Calculation** Question No. 114.
LINE LOAD/ SIZE SERVICE	227.17A /250A	160.58A/175A

220.83(A) - Where Additional Air-Conditioning Equipment or Electric Space-Heating Equipment Is Not to Be Installed (Existing Dwelling Unit)

115. An existing 1560SF home is supplied by a 200A service where the supplying source of voltage is 240/120V-1ϕ. When the house was initially constructed the calculated electrical load resulted to 32,568VA. Since then the homeowners have remodeled their master bedroom, master bathroom and kitchen to include the addition of the following electrical loads:

> 1 - 6.5kW tankless water heater (240V)
> 1 - 7.5kW cooktop (240V)
> 1 - 8.5kW double wall-mounted oven (240V)
> 1 - 3000W wall-mounted heater (240V)
> 1 - 1800W microwave oven (120V)
> 1 - 780VA trash compactor (120V)
> 3 - small-appliances circuits

The home maintained the following existing loads:

> 1 - 240/120V - 5.5kW Dryer
> 1 - 120V Dishwasher - 1260W
> 1 - 120V 560VA garbage disposal
> 3 - 120V 1300W Heat-Vent-Light fixtures
> 2 - AC window units – 240V/15.6A
> 2 - AC window units – 120V/8.2A

Considering all loads, can the existing service be used?

NEC 220.83 provides means for determining whether an existing service or feeder has the capacity to serve additional loads. NEC 220.83 is solely based on the use of either a 240/120V or 208Y/120V, single-phase, 3-wire service.

The provisions of NEC 220.83(A) apply to existing loads that are currently being supplied and the addition of new electrical loads not to include air-conditioning or electric space-heating equipment. NEC 220.83(B) applies to existing electrical loads, added electrical loads and the addition of either air-conditioning or electric space-heating equipment. In summary, the conditions of NEC 220.83(A) and 220.83(B) are very similar to those found in NEC 220.82.

Now let's get started. First we'll determine the existing and added loads.

(1) General lighting and receptacles

$$1560SF \times 3VA = 4680VA$$

(2) Small-appliance (2 required [existing] + 3 added) and laundry

$$1500VA \times 5 = 7500VA$$

(3) Appliances (Actual nameplate rating of appliance)

<u>New</u>
Tankless water heater	6500VA
Cooktop	7500VA
Double wall-mounted oven	8500VA
Wall-mounted heater	3000VA
Microwave oven	1800VA
Trash compactor	780VA

<u>Existing</u>
Dryer	5500VA
Dishwasher	1260VA
Garbage Disposal	560VA
Heat-Vent-Light Fixtures (1300W x 3)	3900VA
AC window units - 240V x 15.6A x 2	7488VA
AC window units - 120V x 8.2A x 2	1968VA
	48,756VA

Total (1) - (3)   = 60,936VA

NEC 220.83(A) requires the first 8kVA of all loads to be calculated at 100 percent and the remainder at 40 percent. Therefore,

1st 8000VA (8kVA)	=	8000VA
Remainder (60,936VA – 8000VA = 52,936VA)		
52,936VA x .40	=	21,174.4VA
		29,174.4VA

Based on the calculated load of 29,174.4VA which results to 121.56A (29,174.4VA /240V) the existing 200A service can be used to supply both existing and added loads.

220.83(B) - Where Additional Air-Conditioning Equipment or Electric Space-Heating
            Equipment Is to Be Installed (Existing Dwelling Unit)

116.   In conjunction with question No. 115., the homeowners decided to add an additional 1375SF to their home; along with having a new air-conditioning and electric heating system installed, which allowed them to remove all existing AC window and wall-mounted heater units. The new system will consist of the following equipment:

> 2 - 4200VA 3 ton A/C unit (compressor and fan motor) (240V)
> 1 - 15kW electric furnace (240V)
> 1 - 10kW electric furnace (240V)

Considering these new loads, can the existing service still be used?

Because new air-conditioning and electric heating equipment will be added to the home, NEC 220.83(B) must be applied.

AC or Heat

AC - 4200VA x 2	=	8400VA
Heat - 15,000W (VA) + 10,000W (VA)	=	25,000VA (Larger)

Other Loads

(1) General lighting and receptacles

$$1560SF + 1375SF \times 3VA = 8805VA$$

(2) Small-appliance and laundry

$$1500VA \times 5 = 7500VA$$

(3) Appliances (Actual nameplate rating of appliance)

<u>New</u>

Tankless water heater	6500VA
Cooktop	7500VA
Double wall-mounted oven	8500VA
Microwave oven	1800VA
Trash compactor	780VA

<u>Existing</u>

Dryer	5500VA
Dishwasher	1260VA
Garbage Disposal	560VA
Heat-Vent-Light Fixtures (1300W x 3)	<u>3900VA</u>
	36,300VA

(4) Total – Other Loads   52,605VA

NEC 220.83(B) requires the AC and Heating loads to be calculated at 100 percent with the larger of the two loads included in the calculation. The first 8kVA of all other loads is calculated at 100 percent and the remainder of all other loads is calculated at 40 percent (.40). Therefore,

(1) Heating Load (100%)   = 25,000VA
(2) Other Loads
     1st 8000VA (8kVA)   =   8000VA
     Remainder (52,605VA – 8000VA = 44,605VA)
     44,605VA x .40   = <u>17,842VA</u>
                                  50,842VA

Based on a calculated load of 50,842VA the load in amperes is 211.84A (50,842VA/240V) which is in excess of the existing 200 amps service. Therefore, as a minimum the service must be upgraded to 225 amps.

220.84(A) - Feeder or Service Load (Multifamily Dwelling) (See conclusion - question No. 119.)

220.84(B) - House Loads (Multifamily Dwelling) (See question No. 87 - Multifamily Dwelling house loads at 208/120V - 3φ and 240/120V - 1φ)

Refer to **(Worksheet D** - Volume **4)** OPTIONAL LOAD CALCULATIONS FOR MULTIFAMILY DWELLING for related questions.

220.84(C)(3) - Connected Loads-Appliances (Multifamily Dwelling) (117. - 120.) [4]

117. A 180 unit multifamily dwelling is equipped with 5.5kW electric dryers. Using the optional calculation per NEC 220.84, determine the service demand load for the dryers.

According to NEC 220.84(C)(3)(c.) the nameplate rating of the appliances (dryers) must be applied to determine the connected load of the appliances.

$$5500 \text{ watts x } 180 = 990,000 \text{ watts}$$

Once this is completed the demand factors of Table 220.84 can be applied per NEC 220.84(C). Based on 180 units a 23 percent (.23) demand factor can be used to determine the service demand load for the electric dryers.

$$990,000 \text{ watts x } .23 = 227,700 \text{ watts (dryers service demand load)}$$

118. What is the computed load in amperes for a 37 unit multifamily dwelling supplied by a 4W, 3φ, 208Y/120V service with a connected load of 1076kVA and a house load of 33kVA?

Based on the provisions of NEC 220.84(C) the given connected load of 1076kVA for the multifamily dwelling units is permitted to be reduced per demand factors of Table 220.84 as follows:

Apply Demand Factors <Table 220.84> For 37 dwelling units the demand factor = 29 percent (.29)

$$(1076\text{kVA}) \ 1,076,000\text{VA x } .29 = 312,040\text{VA} \ (312.04\text{kVA})$$

House Load - To be applied as is.

Total Demand Load

$$
\begin{array}{r}
312,040\text{VA (Multifamily Load)} \\
+ \quad \underline{33,000\text{VA}} \text{ (House Load)} \\
345,040\text{VA (Total Demand Load)}
\end{array}
$$

DWELLING'S OPERATING LINE VOLTAGE - 208V - 3φ
(Given operating voltage or as determined per test examination)

CALCULATE MINIMUM LINE LOAD
(Divide Total Demand Load [VA] by operating line voltage [V])

LINE LOAD = 345,040VA / 208V x 1.732 = 957.76A

119. Re-calculate the service load and determine the size service needed for the multifamily dwelling in question No. 86. using the optional load calculation.

NEC 220.84(C) – Connected Loads (1. – 5.)

1. General Lighting and Receptacle Loads <NEC 220.84(C)(1)>

925SF x 28 x 3VA = 77,700VA

2. Small-Appliances* and Laundry Circuit** Loads <NEC 220.84(C)(2)> - (At least two [2] small-appliance and one [1] branch circuits must be included.)

1500VA x (2* + 1**) x 28 =126,000VA

3. Appliances and Motors Loads <NEC 220.84(C)(3) and (4)> - (Use nameplate rating of each appliance)

Appliances and Motors Loads	VA rating		Number of units		Total VA
Cooktops	5000	x	28	=	140,000
Dishwashers	1200	x	28	=	33,600
Disposals	760	x	28	=	21,280
Dryers	4500	x	28	=	126,000
Microwaves	1000	x	28	=	28,000
Trash Compactors	650	x	28	=	18,200
Wall-mounted ovens	4700	x	28	=	131,600
Water Heaters	5000	x	28	=	140,000
Blowers	560	x	28	=	15,680
HVLs	1500	x	28	=	42,000

Total Appliances Load = 696,360

4. Air-Conditioning (AC) or Fixed Electric Space-Heating (Heat) Load <NEC 220.84(C)(5)> (Larger)

AC = 4700VA
Heat = 10,000W (VA)

AC or Heat	VA rating		Number of units		Total VA
Heat	10,000	x	28	=	280,000

Apply Demand Factors <NEC References - 220.84(A) and Table 220.84>

Total Connected Load = 1,180,060VA
[Loads **1. – 4.** added]

28 units = 33 percent (.33)

1,180,060VA x .33 = 389,419.8VA (Total Demand Load)

DWELLING'S OPERATING LINE VOLTAGE - 240V
(Given operating voltage or as determined per test examination)

CALCULATE MINIMUM LINE LOAD
(Divide Total Demand Load [VA] by operating line voltage [V])

LINE LOAD = 389,419.8VA / 240V = 1622.58A

The service load for this multifamily dwelling using the optional calculation is as calculated per line load.

COMPARISON OF RESULTS

	Standard Calculation Question No. 86.	**Optional Calculation** Question No. 119.
LINE LOAD/ SIZE SERVICE	2763.51A/3000A	1622.58A/2000A

Now consider the service load of the apartment complex in question No. 86. Apply the *Exception* to NEC 220.84(A)(2) where the electric cooktops and wall-mounted cooking units are now replaced with gas appliances.

If the cooktops and wall-mounted cooking units were gas appliances, the line (service) load for the apartment complex based on the standard load calculation-Part III would be reduced to 614,355VA (663,243VA – 48,888VA).

According to the *Exception* to NEC 220.84(A)(2), when a feeder or service load is calculated for a multifamily dwelling per Part III of Article 220 (Standard Load Calculation) where electric cooking appliances are not available and the standard load calculation exceeds an identical load calculation per Part IV (Optional Load Calculation); where the simulated use of electric cooking appliances at 8kW per appliance are included in the calculation, the lesser of the two calculated loads is permitted to be used.

If the loads of the cooktops (140,000VA) and wall-mounted ovens (131,600VA) were removed from item **3**. of question No. 119. and replaced with simulated electric cooking appliances at 8kW per appliance (224,000VA) [8000VA x 28] the total appliance load would then amount to 648,760VA (696,360VA – [140,000VA + 131,600VA] + 224,000VA) instead of 696,360VA. As a result, the total connected load would then become 1,132,460VA (77,700VA + 126,000VA + 648,760VA + 280,000VA). Upon applying the demand factors of Table 220.84, the total demand or service load now becomes 373,711.8VA (1,132,460VA x .33).

In compliance with the *Exception* to NEC 220.84(A)(2), the optional load calculation using simulated cooking appliances yields the lesser of the two loads (373,711.8VA) compared to the results of the standard load calculation where the assumed use of gas appliances are applied (614,355VA). This results to the sole use of the optional load calculation for determining the service load of the multifamily dwelling (apartment complex).

120. Re-calculate the 208/120V, 3-phase service load and determine the size service needed for the 54 unit, the 15 unit, the 17 unit, the 12 unit, and the 10 unit multifamily dwellings in question No. 87. using the optional load calculation. Assume all dryers at 5000W.

| 54 units - 208/120V 3-phase |

1. General Lighting and Receptacle Loads <NEC 220.84(C)(1)>

$$
\begin{array}{llr}
\text{Floor Plan 1 -} & 650\text{SF x 15 x 3VA} = & 29{,}250\text{VA} \\
\text{Floor Plan 2 -} & 900\text{SF x 17 x 3VA} = & 45{,}900\text{VA} \\
\text{Floor Plan 3 -} & 1200\text{SF x 12 x 3VA} = & 43{,}200\text{VA} \\
\text{Floor Plan 4 -} & 1400\text{SF x 10 x 3VA} = & \underline{42{,}000\text{VA}} \\
& \text{Total (Floor Plans)} = & 160{,}350\text{VA}
\end{array}
$$

2. Small-Appliances* and Laundry Circuit Loads** <NEC 220.84(C)(2)> - (At least two [2] small-appliance and one [1] branch circuits must be included)

$$
\begin{array}{lr}
1500\text{VA x 2* x 54} = & 162{,}000\text{VA} \\
1500\text{VA x 1** x 39} = & \underline{58{,}500\text{VA}} \\
& 220{,}500\text{VA}
\end{array}
$$

** 650SF units **(15)** not equipped with washer/dryer connections.

3. Appliances and Motors Loads <NEC 220.82(C)(3) and (4)> - (Use nameplate rating of each appliance)

Appliances and Motors Loads	VA rating		Number of units		Total VA
Cooktops	6300	x	12	=	75,600
Cooktops	6700	x	10	=	67,000
Dishwashers	900	x	32	=	28,800
Dishwashers	1250	x	22	=	27,500
Disposals	600	x	32	=	19,200

Disposals	900	x 22	=	19,800
Dryers	5000	x 39	=	195,000
Microwaves	1000	x 15	=	15,000
Microwaves	1200	x 17	=	20,400
Microwaves	1600	x 12	=	19,200
Microwaves	1800	x 10	=	18,000
Ranges	6100	x 15	=	91,500
Ranges	7800	x 17	=	132,600
Trash compactors	780	x 22	=	17,160
WM Ovens	5400	x 12	=	64,800
WM Ovens	5800	x 10	=	58,000
Water Heaters	3380	x 54	=	182,520
Blowers	1044	x 15	=	15,660
Blowers	1176	x 17	=	19,992
Blowers	1320	x 12	=	15,840
Blowers	1416	x 10	=	14,160
Heat-Vent-Lights	1300	x 15	=	19,500
Heat-Vent-Lights	1600	x 29	=	46,400
Heat-Vent-Lights	1750	x 10	=	17,500

Total Appliances Load = 1,201,132

4. Air-Conditioning (AC) or Fixed Electric Space-Heating (Heat) Load <NEC 220.84(C)(5)> (Larger)

AC /Heat	VA rating		Number of units	Total VA
AC	3648VA	x	15	= 54,720
AC	4560VA	x	17	= 77,520
AC	5472VA	x	12	= 65,664
AC	6384VA	x	10	= 63,840
				261,744
Heat	10,000	x	15	= 150,000
Heat	12,000	x	17	= 204,000
Heat	15,000	x	12	= 180,000
Heat	18,000	x	10	= 180,000
				714,000 (Larger)

Apply Demand Factors <NEC References - 220.84(A) and Table 220.84>

Total Connected Load = 2,295,982VA
[Loads 1. – 4. added]
54 units = 25 percent (.25)
2,295,982VA x .25 = 573,995.5VA (Total Demand Load)

DWELLING'S OPERATING LINE VOLTAGE - 208V - 3ϕ
(Given operating voltage or as determined per test examination)

CALCULATE MINIMUM LINE LOAD
(Divide Total Demand Load [VA] by operating line voltage [V])

$$\text{LINE LOAD} = \underline{573{,}995.5}\text{VA} / \underline{208\text{V x }1.732}\text{V} = \mathbf{1593.3A}$$

The service load for this multifamily dwelling using the optional calculation is as calculated per line load.

COMPARISON OF RESULTS

	Standard Calculation Question No. 87.	**Optional Calculation** Question No. 120.
LINE LOAD/ SIZE SERVICE	3805.4A/4000A	1593.3.88A/1600A

650SF units (15) - 208/120V 3-phase

1. General Lighting and Receptacle Loads <NEC 220.84(C)(1)>

$$650\text{SF x }15\text{ x }3\text{VA} = 29{,}250\text{VA}$$

2. Small-Appliances* and Laundry Circuit** Loads <NEC 220.84(C)(2)> - (At least two [2] small-appliance and one [1] branch circuits must be included)

$$1500\text{VA x }(2^* + 0^{**})\text{ x }15 = 45{,}000\text{VA}$$
** 650SF units **(15)** not equipped with washer/dryer connections.

3. Appliances and Motors Loads <NEC 220.82(C)(3) and (4)> - (Use nameplate rating of each appliance)

Appliances and Motors Loads	VA rating		Number of units		Total VA
Dishwashers	900	x	15	=	13,500
Disposals	600	x	15	=	9000
Microwaves	1000	x	15	=	15,000
Ranges	6100	x	15	=	91,500
Water Heaters	3380	x	15	=	50,700
Blowers	1044	x	15	=	15,660
Heat-Vent-Lights	1300	x	15	=	19,500
		Total Appliances Load	=		214,860

4. Air-Conditioning (AC) or Fixed Electric Space-Heating (Heat) Load <NEC 220.84(C)(5)> (Larger)

AC/Heat	VA rating		Number of units		Total VA
AC	3648	x	15	=	54,720
Heat	10,000	x	15	=	150,000(larger)

Apply Demand Factors <NEC References - 220.84(A) and Table 220.84>

Total Connected Load = 439,110VA

[Loads **1.** - **4.** added]

15 units = 40 percent (.40)

439,110VA x .40 = 175,644VA (Total Demand Load)

DWELLING'S OPERATING LINE VOLTAGE - 208V - 3φ
(Given operating voltage or as determined per test examination)

CALCULATE MINIMUM LINE LOAD
(Divide Total Demand Load [**VA**] by operating line voltage [**V**])

LINE LOAD = 175,644VA / 208V x 1.732 = **487.6A**

The service load for this multifamily dwelling using the optional calculation is as calculated per line load.

COMPARISON OF RESULTS

	Standard Calculation Question No. 87.	**Optional Calculation** Question No. 120.
LINE LOAD/ SIZE SERVICE	848.68A/1000A	487.6A/500A

900SF units (17) - 208/120V 3-phase

1. General Lighting and Receptacle Loads <NEC 220.84(C)(1)>

900SF x 17 x 3VA = 45,900VA

2. Small-Appliances* and Laundry Circuit** Loads <NEC 220.84(C)(2)> - (At least two [2] small-appliance and one [1] branch circuits must be included)

1500VA x (2* + 1**) x 17 = 76,500VA

3. Appliances and Motors Loads <NEC 220.82(C)(3) and (4)> - (Use nameplate rating of each appliance)

Appliances and Motors Loads	VA rating		Number of units		Total VA
Dishwashers	900	x	17	=	15,300
Disposals	600	x	17	=	10,200
Dryers	5000	x	17	=	85,000
Microwaves	1200	x	17	=	20,400
Ranges	7800	x	17	=	132,600
Water Heaters	3380	x	17	=	57,460
Blowers	1176	x	17	=	19,992
Heat-Vent-Lights	1600	x	17	=	27,200

Total Appliances Load = 368,152

4. Air-Conditioning (AC) or Fixed Electric Space-Heating (Heat) Load <NEC 220.84(C)(5)> (Larger)

AC or Heat	VA rating		Number of units		Total VA
AC	4560	x	17	=	77,520
Heat	12,000	x	17	=	204,000 (Larger)

Apply Demand Factors <NEC References - 220.84(A) and Table 220.84>

Total Connected Load = 694,552VA
[Loads **1.** - **4.** added]
17 units = 39 percent (.39)
694,552VA x .39 = 270,875.28VA (Total Demand Load)

DWELLING'S OPERATING LINE VOLTAGE - 208V - 3φ
(Given operating voltage or as determined per test examination)

CALCULATE MINIMUM LINE LOAD
(Divide Total Demand Load [VA] by operating line voltage [V])

LINE LOAD = 270,875.28VA / 208V x 1.732 = **751.9A**

The service load for this multifamily dwelling using the optional calculation is as calculated per line load.

COMPARISON OF RESULTS

	Standard Calculation Question No. 87.	**Optional Calculation** Question No. 120.
LINE LOAD/ SIZE SERVICE	1245.38/1600A	751.9A/800A

1200SF units (12) - 208/120V 3-phase

1. General Lighting and Receptacle Loads <NEC 220.84(C)(1)>

$$1200SF \times 12 \times 3VA = 43,200VA$$

2. Small-Appliances* and Laundry Circuit** Loads <NEC 220.84(C)(2)> - (At least two [2] small-appliance and one [1] branch circuits must be included)

$$1500VA \times (2^* + 1^{**}) \times 12 = 54,000VA$$

3. Appliances and Motors Loads <NEC 220.82(C)(3) and (4)> - (Use nameplate rating of each appliance)

Appliances and Motors Loads	VA rating		Number of units		Total VA
Cooktops	6300	x	12	=	75,600
Dishwashers	1250	x	12	=	15,000
Disposals	900	x	12	=	10,800
Dryers	5000	x	12	=	60,000
Microwaves	1600	x	12	=	19,200
Trash compactors	780	x	12	=	9360
WM Ovens	5400	x	12	=	64,800
Water Heaters	3380	x	12	=	40,560
Blowers	1320	x	12	=	15,840
Heat-Vent-Lights	1600	x	12	=	19,200

Total Appliances Load = 330,360

4. Air-Conditioning (AC) or Fixed Electric Space-Heating (Heat) Load <NEC 220.84(C)(5)> (Larger)

AC or Heat	VA rating		Number of units		Total VA
AC	5472	x	12	=	65,664
Heat	15,000	x	12	=	180,000 (Larger)

Apply Demand Factors <NEC References - 220.84(A) and Table 220.84>

Total Connected Load = 607,560VA
[Loads **1. - 4.** added
12 units = 41 percent (.41)
607,560VA x .41 = 249,099.6VA (Total Demand Load)

DWELLING'S OPERATING LINE VOLTAGE - 208V - 3ϕ
(Given operating voltage or as determined per test examination)

CALCULATE MINIMUM LINE LOAD

(Divide Total Demand Load [VA] by operating line voltage [V])

LINE LOAD = 249,099.6VA / 208V x 1.732 = **691.45A**

The service load for this multifamily dwelling using the optional calculation is as calculated per line load.

COMPARISON OF RESULTS

	Standard Calculation Question No. 87.	**Optional Calculation** Question No. 120.
LINE LOAD/ SIZE SERVICE	1082.91A/1200A	691.45A/700A

1400SF units (10) - 208/120V 3-phase

1. General Lighting and Receptacle Loads <NEC 220.84(C)(1)>

1400SF x 10 x 3VA = 42,000VA

2. Small -Appliances* and Laundry Circuit** Loads <NEC 220.84(C)(2)> - (At least two [2] Small-appliance and one [1] branch circuits must be included)

1500VA x (2* + 1**) x 10 = 45,000VA

3. Appliances and Motors Loads <NEC 220.82(C)(3) and (4)> - (Use nameplate rating of each appliance)

Appliances and Motors Loads	VA rating		Number of units		Total VA
Cooktops	6700	x	10	=	67,000
Dishwashers	1250	x	10	=	12,500
Disposals	900	x	10	=	9000
Dryers	5000	x	10	=	50,000
Microwaves	1800	x	10	=	18,000
Trash compactors	780	x	10	=	7800
WM Ovens	5800	x	10	=	58,000
Water Heaters	3380	x	10	=	33,800
Blowers	1416	x	10	=	14,160
Heat-Vent-Lights	1750	x	10	=	17,500

Total Appliances Load = 287,760

4. Air-Conditioning (AC) or Fixed Electric Space-Heating (Heat) Load<NEC 220.84(C)(5)> (Larger)

AC or Heat	VA rating		Number of units		Total VA
AC	6384	x	10	=	63,840
Heat	18,000	x	10	=	180,000 (Larger)

Apply Demand Factors <NEC References - 220.84(A) and Table 220.84>

Total Connected Load = 554,760VA
[Loads **1.** - **4.** added]
10 units = 43 percent (.43)
554,760VA x .43 = 238,546.8VA (Total Demand Load)

DWELLING'S OPERATING LINE VOLTAGE - 208V - 3ϕ
(Given operating voltage or as determined per test examination)

CALCULATE MINIMUM LINE LOAD
(Divide Total Demand Load [VA] by operating line voltage [V])

LINE LOAD = 238,546.8VA / 208V x 1.732 = **662.16A**

The service load for this multifamily dwelling using the optional calculation is as calculated per line load.

COMPARISON OF RESULTS

	Standard Calculation Question No. 87.	**Optional Calculation** Question No. 120.
LINE LOAD/ SIZE SERVICE	1031.86A/1200A	662.16A/800A

220.85 - Two Dwelling Units

121. A duplex apartment is supplied from the same 240/120V service feeder. Both units of the duplex are 840SF and consist of the following loads:

240/120V
7.5kW range
4.5kW dryer

120V
740W dishwasher
1500W HVL
960VA disposal

240V
3.38kW water heater
10kW electric heat
AC - 11.3A compressor, 1.5A fan motor

— 282 —

Use the standard calculation to determine the service load of the duplex apartment.

1. General Lighting and Receptacle Loads <NEC 220.12, Table 220.12, 220.14(J) and 220.42> (Open porches, garages, unused or unfinished spaces not adaptable for future use not included)

$$840SF \times 3VA \times 2 = 5040VA$$

2. Small-Appliance Circuit Load (Portable) <NEC 220.52(A)>

$$1500VA \times 2 \times 2 = 6000VA$$

3. Laundry Circuit Load <NEC 220.52(B)>

$$1500VA \times 2 = \underline{3000VA}$$

TOTAL (Lines 1. – 3.) = 14,040VA
(If Total VA is less than or equal to 120,000VA, step c. is not required)

APPLY DEMAND FACTORS <NEC and Table 220.42>
a. First 3000VA of above TOTAL (At 100%) =   3000VA
b. 11,040 VA   x .35   =   3864VA
   (Total VA – 3001 VA up to 117,000VA)
c.   =   0
       TOTAL (Lines a. – c.)   =   6864VA

GENERAL LIGHTING and RECEPTACLE, SMALL-APPLIANCE AND LAUNDRY LOADS

1. - 3. LINE LOAD

    6864VA

4. Appliance Loads (Fastened-In-Place) <NEC 220.53> (Use nameplate rating of each appliance. Electric ranges, dryers, space-heating equipment or air-conditioning equipment not included)

Appliances

1. Dishwashers - 740VA x 2   =   1480VA
2. HVLs - 1500VA x 2   =   3000VA
3. Disposals - 960VA x 2   =   1920VA
4. Water Heaters - 3380VA x 2 =   <u>6760VA</u>

    APPLIANCES TOTAL = 13,160VA

APPLY DEMAND FACTOR (Applicable, when number of above appliances exceeds four (4) or more)

(Appliances Total) 13,160VA x .75 = 9870VA

APPLIANCE LOADS

4.   LINE LOAD

9870VA

5. Clothes Dryers <NEC and Table 220.54> (Use 5000W [VA] or nameplate rating, whichever is larger)

CLOTHES DRYERS

5.   LINE LOAD

10,000VA

6. Electric Ranges and Other Cooking Appliances <NEC and Table 220.55>

Minimum Demand Load determined per Columns B and C.

Applying Column B - 7.5kW x 2 x .65 = 9.75kW *(Use minimum demand load per Column B)*
Applying Column C - 2 appliances under 12kW – Maximum demand = 11kW

RANGE/COOKING APPLIANCES

6.   LINE LOAD

9750VA

7. Heating and Air-Conditioning (AC) Equipment <NEC References - 220.50, 220.51, 220.60, Table 430. 248 and 440.6(A)> (Include VA rating of air handler [blower].)

Heating Load

    Electric Heat
    10kW x 2 = 20kW (20,000VA) (Total Heat)

AC Load

        Compressor(s)   Fan Motor(s)
    [(240V x 11.3A) + (240V x 1.5A)] x 2 = 6144VA (Total AC)

Heating load is larger.

7.   <u>LINE LOAD</u>

   20,000VA

8. Largest Motor <NEC 430.17, 430.24, 440.7 and 440.33> (Use motor with highest full-load current (FLC) regardless of voltage rating)

Largest Motor (LM) = 240V x 11.3A = 2712VA (AC compressor)

<u>2712VA</u> x .25 = 678VA
(LM)

LARGEST MOTOR

8.   <u>LINE LOAD</u>

   678VA

TOTAL DEMAND LOAD (LINE AND NEUTRAL) (Add lines 1. - 8.)

   <u>LINE LOAD</u>

1. – 3.   6864VA
4.   9870VA
5.   10,000VA
6.   9750VA
7.   20,000VA
8.   <u>678VA</u>
      57,162VA

DWELLING'S OPERATING LINE VOLTAGE - <u>240V</u>
(Given operating voltage or as determined per test examination)

LINE LOAD = <u>57,162</u>VA / <u>240</u>V = **238.18A**

According to NEC 220.85, when two dwelling units are supplied by a single feeder and the calculated load under Part III (Standard Load Calculation) of Article 220 exceeds that for three identical units calculated under NEC 220.84, the lesser of the two loads shall be permitted to be used.

Having performed the standard load calculation for the duplex apartment the next thing is to apply the optional load calculation considering three identical units of the duplex apartment instead of two at the same square footage and electrical loads.

1. General Lighting and Receptacle Loads <NEC 220.84(C)(1)>

   840SF x 3 x 3VA = 7560VA

2. Small-Appliances* and Laundry Circuit** Loads <NEC 220.84(C)(2)> - (At least two [2] small-appliance and one [1] branch circuits must be included)

$$1500VA \times (2^* + 1^{**}) \times 3 = 13,500VA$$

3. Appliances and Motors Loads <NEC 220.82(C)(3) and (4)> - (Use nameplate rating of each appliance)

Appliances and Motors Loads	VA rating		Number of units		Total VA
Dishwashers	740	x	3	=	2220
Disposals	960	x	3	=	2880
Dryers	4500	x	3	=	13,500
Ranges	7500	x	3	=	22,500
Water Heaters	3380	x	3	=	10,140
Heat-Vent-Lights	1500	x	3	=	4500

$$\text{Total Appliances Load} = 55,740$$

4. Air-Conditioning (AC) or Fixed Electric Space-Heating (Heat) Load <NEC 220.84(C)(5)> (Larger)

AC $= 240V \times (11.3A + 1.5A) = 3072VA$
Heat $= 10,000W$ (VA)

AC/Heat	VA rating		Number of units		Total VA
AC	3072	x	3	=	9216
Heat	10,000	x	3	=	30,000

Apply Demand Factors <**NEC References** - 220.84(A) and Table 220.84>

Total Connected Load = 106,800VA
[Loads **1. – 4.** added]
3 units = 45 percent (.45)
106,800VA x .45 = 48,060VA (Total Demand Load)

DWELLING'S OPERATING LINE VOLTAGE - 240V
(Given operating voltage or as determined per test examination)

CALCULATE MINIMUM LINE LOAD
(Divide Total Demand Load [**VA**] by operating line voltage [**V**])

LINE LOAD = 48,060VA / 240V = **200.25A**

Applying the standard load calculation for the duplex apartment, the line load was calculated at 238.1A compared to 200.25A as the results when the multifamily optional load calculation was applied using 3 identical units. Based on the provisions of NEC 220.85, the lesser of the two loads is permitted to be used. Therefore, the single feeder for the duplex apartment can be determined based on 200.25A.

220.86 - Optional Load Calculation for Schools (122. - 123.) [2]

122. Using the optional load calculation, what size parallel service conductors (4 per phase) are required for a school having a service demand of 11.45VA/SF at 130,000SF. The school is supplied by a 480/277V, 3-phase service. Consider the use of copper conductors at 75°C.

Per Table 220.86

1st 3VA/SF (at 100%)	- 3VA/SF x 130,000SF	=	390,000VA
Over 3VA/SF up to 20VA/SF (at 75%)	- 8.45VA/SF x 130,000SF x .75	=	823,875VA
			1,213,875VA

$$1,213,875VA / 480V \times 1.732 = 1460.11A$$

Assuming the use of a 1600A service (based on 1460.11A) where 1600A/4 = 400A. Use four 600 kcmil copper conductors which are rated for 420A.

123. Refer to the community college in question No. 101. Based on the information provided in the standard load calculation for the school, re-calculate the load using the optional calculation.

According to NEC 220.86, the connected load to which the demand factors of Table 220.86 apply shall include all of the interior and exterior lighting, power, water heating, cooking, other loads, and the larger of the air-conditioning or space heating load with the building or structure. In this particular case as it pertains to the community college in question No. 101. all requirements were met.

Using the square footage (495,000SF) and the connected load (26,639,059.25VA* or 26.639MVA [rounded-off]) of the community college (at 100 percent) the demand factors of Table 220.86 can be applied to determine the service load of the school's entire electrical system per optional calculation based on the ratio of the actual load (VA) with respect to the given square feet (SF). Applying both units the connected load results to 53.82VA/SF (26.639MVA /495,000SF).

* **(1)** 1,856,250.00VA + **(2)** 1,547,493.13VA + **(3)** 2,466,140.00VA + **(4)** 657,575.75VA + **(5)** 2,548,774.30VA + **(6)** 4,937,227.67VA + **(9)** 12,599,826.2VA + **(10)** 25,772.2VA = 26,639,059.25VA (connected load, question 101.)

Per Table 220.86

1st 3VA/SF (at 100%)	- 3VA/SF x 495,000SF	=	1,485,000.0VA
Over 3VA/SF up to 20VA/SF (at 75%)	- 17VA/SF x 495,000SF x .75	=	6,311,250.0VA
Over 20VA/SF (at 25%)	- 33.82VA/SF x 495,000SF x .25	=	4,185,225.0VA
			11,981,475.0VA

Compared to the school's service load of 3697.24A (26,639,059.25VA / 4160V x 1.732) per standard load calculation in question No. 101., the optional load calculation for the same school yields a service load of 1662.91A (11,981,475.0VA / 4160V x 1.732).

Unlike a feeder or service (line) load, NEC 220.86 makes no provisions for reducing the neutral load based on an optional load calculation. Therefore, when applying the optional load calculation for a school, the neutral load is always based on the provisions of Part III (standard calculation) of Article 220 where in question 101, the service neutral load was calculated at 733.80A.

220.87 - Determining Existing Load

124. An existing 3-phase service is supplied by four-350 kcmil THWN copper conductors from a 480/277V source. After one year of continuous monitoring it was noted that the service load peaked at 174.6kVA. If needed, how much additional load can be added to the service?

Considering that the first (1) condition of NEC 220.87 has been met, the next approach will be to ensure that the second (2) condition is also met. Multiplying the maximum or peak demand (174.6kVA) by 125 percent [per condition (2)] produces an increased load of 218.25kVA. When converted to amperes, the increased load yields 262.52 amps (218.25kVA / 480V x 1.732). Based on the conclusion of condition (2), if an additional load was added to the increased load the combination could not exceed the ampacity of the 350 kcmil copper service conductors or the rating of the service. Therefore, if the increased load (amps) was subtracted from the rated ampacity of the service conductors the difference would produce the load that could be added to the service. Table 310.15(B)(16) lists the rated ampacity of a 350 kcmil THWN copper conductor at 310A. As a result (310A – 262.52A = 47.48A), an approximate load of 39.47kVA (480V x 47.48A x 1.732) can be added to the existing 174.6kVA service load. As for condition three (3), in conjunction with condition (2) a 350A overcurrent device (rating of the service) could be used based on the use of 350 kcmil conductors (310A) and be in compliance with NEC 240.4[240.4(B)] and NEC 230.90.

220.88 - Optional Load Calculations for New Restaurant (125. – 127.) [3]

125. What is the calculated load in amperes for a new, "All Electric Restaurant", served by a 3-phase, 4-wire, 240/120V service, with a connected load of 685kVA?

Since the connected load is 685kVA, the formula found in row 3 of Column 2 in Table 220.88 is referenced to calculate the load in amperes for this new "All Electric Restaurant". The formula to determine the optional load calculation for the restaurant requires 50 percent (.50) of the amount over 325kVA to be added to 172.5kVA. Applying the formula,

$$\begin{aligned} \text{Load} &= (685\text{kVA} - 325\text{kVA}) \times .50 + 172.5\text{kVA} \\ &= (360\text{kVA}) \times .50 + 172.5\text{kVA} \\ &= 180\text{kVA} + 172.5\text{kVA} \\ &= 352.5\text{kVA} \end{aligned}$$

The load in amperes for the new "All Electric Restaurant" is 848A (352.5kVA / 240V x 1.732).

126. Considering the use of the same connected load (685kVA) in question No. 125., what is the calculated load in amperes for a new, "Not All Electric Restaurant", served by a 3-phase, 4-wire 240/120V service?

Applying the connected load, the formula found in row 3 of Column 3 in Table 220.88 is referenced to calculate the load in amperes for this new "Not All Electric Restaurant". The formula to determine the optional load calculation for the restaurant requires 45 percent (.45) of the amount over 325kVA to be added to 262.5kVA. Applying the formula,

$$\begin{aligned} \text{Load} &= (685\text{kVA} - 325\text{kVA}) \times .45 + 262.5\text{kVA} \\ &= (360\text{kVA}) \times .45 + 262.5\text{kVA} \\ &= 162\text{kVA} + 262.5\text{kVA} \\ &= 424.5\text{kVA} \end{aligned}$$

The load in amperes for the new "Not All Electric Restaurant" is 1021.22A (424.5kVA / 240V x 1.732).

127. Using the information in question No. 100., apply the optional load calculation for the new restaurant based on the total connected load of 172,503.38VA (172.5kVA [rounded-off]) to determine the service load of the restaurant in amperes.

Applying the connected load, the formula found in row 1 of Column 2 in Table 220.88 is referenced to calculate the load in amperes for this new "All Electric Restaurant". The formula to determine the optional load calculation for the restaurant requires 80 percent (.80) of the connected load. Applying the formula,172,503.38VA x .80 = 138,002.7VA.

The load in amperes for the new "All Electric Restaurant" is 383.07A (138,002.7kVA / 208V x 1.732).

Compared to the restaurant's service load of 478.84A per standard load calculation in question No. 100., the optional load calculation for the same restaurant yields a service load of 383.07A.

Re-calculate the service load of the restaurant in amperes if the connected load was 1,172,503.38VA (1172.50kVA [rounded-off]).

Applying the new connected load, the formula found in row 4 of Column 2 in Table 220.88 is referenced to calculate the load in amperes based upon this new requirement. The formula to determine the optional load calculation requires 50 percent (.50) of the amount over 800kVA to be added to 410kVA. Applying the formula,

$$\begin{aligned} \text{Load} &= (1172.50\text{kVA} - 800\text{kVA}) \times .50 + 410\text{kVA} \\ &= (372.5\text{kVA}) \times .50 + 410\text{kVA} \\ &= 186.25\text{kVA} + 410\text{kVA} \\ &= 596.25\text{kVA} \end{aligned}$$

The new load in amperes for the restaurant is 1655.07A (596.25kVA / 208V x 1.732).

PART V - Farm Load Calculations

The following procedures for calculating feeder or service loads for farms are based on applicable sections of Part V of Article 220 as referenced.

220.102(A) - Dwelling Unit (Farm Loads - Buildings and Other Loads)

Whether applying the standard load (Part III) or the optional load (Part IV) calculations of Article 220, the same procedures are applied when calculating a feeder or service load for a farm dwelling. However, when a farm dwelling and farm loads are supplied by a common service and the dwelling has electric heat along with the farm having an electric grain-drying system the optional load calculation per Part IV can (shall) not be used. Because both procedures (standard and optional) have been extensively covered in other examples of this article, the need to perform a load calculation for a farm dwelling is not necessary.

220.102(B) - Other Than Dwelling Unit (Farm Loads - Buildings and Other Loads)
(128. - 129.) [2]

NEC 220.102(B) is applied when calculating feeder or service farm loads *without* including the electrical loads of an existing farm house (dwelling). According to NEC 220.102(B), where a feeder or service supplies a farm load [building(s) or other load(s) having two or more separate branch circuits] the load for feeders, service conductors, and service equipment shall be calculated in accordance with demand factors not less than indicated in Table 220.102.

Of all provisions of the NEC where load calculations are required, farm loads are the only type load calculations that are required to be performed applying units of amperes (A) opposed to volt-amperes (VA). When performing farm load calculations all loads are calculated at a maximum of 240-volts. As far providing demand factors for determining the size of service neutral conductors for farm loads, there are none thus leaving the neutral conductor to be sized at 100 percent of the anticipated neutral loads. Therefore, where 120-volt loads are involved they must be converted to 240-volt loads simply by doubling the load voltage and reducing the load current by one-half. Where a farm building or other farm loads are supplied by a common service the demand factors of Table 220.102 must be applied. Because interpreting the provisions of Table 220.102 ("as is") may prove to be a bit perplexed, the following breakdown is provided. Table 220.102 is divided into three parts where demand factors of 100 percent, 50 percent, and 25 percent are applied. At 100 percent, the largest of three given loads is selected, be it, **(1)** all loads that are expected to operate simultaneously [at the same time], *or* **(2)** 125 percent of the full-load current of the largest motor, *or* **(3)** the first 60A of the total load. After selecting the largest of the three load options, the next 60A of the total load is then calculated at 50 percent (.50) followed by reducing all remaining loads of the total load by 25 percent (.25).

To better understand the provisions of NEC and Table 220.102 refer to question Nos. 128. and 129.

128. A dairy farm building will be supplied from a 240/120V single-phase source to supply the following loads:

120V
Lighting (continuous) - 74A
Receptacles - 45A

240V
Largest motor, 7.5HP - 40A (Table 430.248)
Other motor loads - 66.3A
Heating/AC - 56A

If 35 percent of the total load will operate at the same time, what size service is required for this building?

The first approach in deriving the service for this dairy farm building is to calculate the total given loads in amperes at 240V per Table 220.102 which states "Ampere Load at 240 Volts Maximum." As a result, the 120 volts-amperes loads are converted to 240 volts-amperes loads which yield half the amperes for the lighting (37A @ 240V) and receptacle (22.5A @ 240V) loads. When all loads are totaled at 240 volts the results are as follows:

$$
\begin{array}{lll}
\text{Lighting - 37.0A x 1.25} & = & 46.25\text{A} \\
\text{Receptacles} & = & 22.50\text{A} \\
\text{Largest motor - 40.0A} & = & 40.00\text{A} \\
\text{Other motor loads} & = & 66.30\text{A} \\
\text{Heating/AC -} & = & \underline{56.00\text{A}} \\
& & 231.05\text{A}
\end{array}
$$

The next approach must be based on the requirements of Table 220.102 where the largest of the simultaneous loads, the largest motor load at 125 percent or the first 60 amperes of the total load is determined and calculated at 100 percent.

1. Simultaneous Load - 231.05A x .35 = **80.87A**

2. Largest motor - 40.0A x 1.25 = 50A

3. First 60 amperes of total load = 60A

In evaluating the three type loads the *simultaneous load* is largest and calculated at 100 percent thus excluding the other two loads.

Afterward, Table 220.102 requires the *next 60 amperes of all other loads* to be calculated at 50 percent.

60A x .50 = **30A**

Although the next 60 amperes of all other loads were reduced by 50 percent, the total load is nevertheless reduced by 80.87A and 60A resulting to the remainder of other loads.

231.05A − 80.87A − 60A = 90.18A

Finally, Table 220.102 requires the *remainder of all other loads* to be calculated at 25 percent. Therefore,

$$90.18A \times .25 = \boxed{\textbf{22.55A}}$$

and when totaled according to the demand factors of Table 220.102 amounts to,

$$80.87A + 30A + 22.55A = \textbf{133.42A}$$

At a calculated load of 133.42A, as a minimum a 150A service is required.

129. A single-family farm home, two farm buildings and several types of farm equipment are supplied by a 400A, three-phase, 4W-240/120V service. A three-phase, 4W-240/120V feeder will be used to support the operation of the farm equipment. Determine the minimum size feeder needed based on the following loads:

 1 - 40HP, 240V, 3P Motor   1 - 10HP, 240V, 1P Motor   1 - 5HP, 120V Motor
 1 - 240V, 3P - 50kVA Motor-Generator Arc Welder (DC @ .80)
 18 - 120V duplex receptacles
 3 - 240V, 3kW Heaters
 28.7A - Continuous 120V Lighting (Simultaneous Use)

To begin, the load amperes of each piece of equipment is identified with all 120V loads being converted to 240V per Table 220.102 and totaled to derive the total load of all given farm equipment.

40HP, 240V, 3P Motor	= 104A [T430.250]
10HP, 240V, 1P Motor	= 50A [T430.248]
5HP, 120V Motor (56A [T430.248] /2)	= 28A @ 240V
240V, 3P – 50kVA Motor-Generator Arc Welder (DC @ .80 (50kVA/240V x 1.732) x .91 [T630.11(A)]	= 109.46A
120V Duplex receptacles (180VA [NEC 220.14(I)] x 18 / 240V)	= 13.5A @ 240V
3kW Heaters [9kW / 240V]	= 37.5A
120V Lighting [28.7A x 1.25 [NEC 215.2(A)(1)] / 2]	= 17.94A @ 240V

$$\text{TOTAL CONNECTED LOAD} \quad = 360.4A \, (360A)$$

Just as question No. 128., the next approach is based on the requirements of Table 220.102 where the largest of the simultaneous loads, the largest motor load at 125 percent *or* the first 60 amperes of the total load is determined and calculated at 100 percent.

1. Simultaneous Load - (Lighting) = 17.94A

2. Largest motor - (40HP, 240V, 3P Motor) - 104A x 1.25 = $\boxed{\textbf{130A}}$

3. First 60 amperes of total load = 60A

In evaluating the three type loads the *largest motor* is largest and calculated at 100 percent thus excluding the other two loads.

Afterward, Table 220.102 requires the *next 60 amperes of all other loads* to be calculated at 50 percent.

$$60A \times .50 = \boxed{30A}$$

Although the next 60 amperes of all other loads were reduced by 50 percent, the total load is reduced by 130A and 60A resulting to the remainder of other loads.

$$360A - 130A - 60A = 170A$$

Finally, Table 220.102 requires the *remainder of all other loads* to be calculated at 25 percent. Therefore,

$$170A \times .25 = \boxed{42.5A}$$

$$130A + 30A + 42.5A = \mathbf{202.5A}$$

At a calculated load of 202.5A, as a minimum a 225A service is required.

220.103 - Farm Loads (Totals) (130. - 131.) [2]

NEC 220.103 is an extension of NEC 220.102. In NEC 220.102 where one service supplies a farm dwelling and a separate service supplies power to farm equipment buildings the provisions of NEC 220.102(A) and 220.102(B) are applied accordingly. Per NEC 220.103 when a farm dwelling and farm equipment building(s) are supplied by a common service, the total load of a farm is used to calculate the minimum size service conductors and service equipment needed where the load of the dwelling unit and the demand factors listed in Table 220.103 are applied.

To better understand the provisions of NEC and Table 220.103 refer to question Nos. 130. and 131.

130. Determine the minimum size 240/120V-1ϕ service needed to supply a two-story farm house and the following farm loads:

Receptacles - 42.8A
Simultaneous - 44.2A

Motors*		**Lighting**
240V	120V	240V - 17.4A
7½ HP - 40A	1HP - 16A	120V - 45.8A
5HP - 28A	¾HP (3) - 13.8A	
3HP - 17A		
* Refer to Table 430.248		

The total load of the farm house is 167.7A.

Before getting started notice how all 120V loads were converted to 240V per Table 220.102.

Motors	**Lighting**	**Receptacles**	**Simultaneous**
240V	240V - 17.4A	42.8A	44.2A
7½ HP - 40A (largest)	120V - 45.8A / 2 = **22.9A @ 240V**		
5HP - 28A			
3HP - 17A			
120V			
1HP - 16A / 2 = **8A @ 240V**			
¾HP (3) - 13.8A / 2 (6.9A) x 3 = **20.7A @ 240V**			

Per NEC 220.102 the farm dwelling unit and demand factors specified in Table 220.103 can now be applied.

Since the load designated as "**simultaneous**" is the largest load, it is calculated at 100 percent followed by the application of the demand factors per individual loads as outlined in Table 220.103. As a reminder, since the simultaneous load is larger than the largest motor load, the largest motor load is not increased by 1.25 percent.

Largest Load @ 100% (Simultaneous) -	44.2A
Second largest load @ 75% (Receptacles) - 42.8A x .75 =	32.1A
Third largest load @ 65% (7.5HP Motor) - 40A x .65 =	26.0A
Remaining loads @ 50% (114A**) - 114A x .50 =	57.0A
	159.3A (Total per Table 220.103)
Farm house -	167.7A
	327.0A (total load)

Motors - 28A + 17A + 8A + 20.7A = 73.7A
Lighting - 17.4A + 22.9   = 40.3A
                     114.0A**

Because the 327A total load includes farm loads and not exclusively a farm dwelling load, Table 310.15(B)(7) cannot be applied to size the service needed to supply the total load. Based on the results of the calculated load the needed service must be sized according to NEC 240.6(A), 408.30 and 408.36. In accordance with NEC 240.6(A) as a minimum a 350A service is needed per main overcurrent device which exceeds the 327A farm load.

131. A dairy, poultry and pig farm utilizes 9 buildings to meet the ongoing demands of supplying four large food-chains. Determine the minimum size service needed to supply electrical power to the buildings and a farm house load calculated at 188.6A. The building loads are as listed:

Building 1 - 133A	Building 2 - 113A	Building 3 - 68.6A
Building 4 - 131.7A	Building 5 - 92.4A	Building 6 - 95A
Building 7 - 55A	Building 8 - 108.8A	Building 9 - 157.2A

NEC 220.103 states where there is equipment in two or more farm equipment buildings or for loads having the same function, such loads shall be calculated in accordance with Table 220.102

and shall be permitted to be combined as a single load in Table 220.103 for calculating the total load.

Considering all loads from the largest to the total remaining loads, first followed by the addition of the farm house load, the calculations are as follows:

Largest Load @ 100% (Building 9) -   157.20A
Second largest load @ 75% (Building 1) - 133A x .75 =   99.75A
Third largest load @ 65% ((Building 4) - 131.7A x .65 =   85.61A
Remaining loads @ 50% (532.8A*) - 532.8A x .50 =   266.40A
                                            608.96A (Total per Table 220.103)
Farm house -   188.60A
                                            797.56A (total load)

   *Remaining building's total load - 55A + 113A + 92.4A + 108.8A + 68.6A + 95A = 532.8A

As a minimum, based on the total calculated load an 800A service is required. Again, refer to NEC 240.6(A), 408.30 and 408.36.

ABOUT THE AUTHOR

For over thirteen years, Alvin Walker, a native of Shreveport, Louisiana owned and operated a small yet successful electrical contracting business. He now works as an author and instructor specializing in electrical and NEC training where his services are available throughout the United States. In his over thirty year of experience, he has developed a very strong background in electrical maintenance, construction, and design engineering. He has taught Business and Law for Contractors, the National Electrical Code, Electrical Theory, (Electrical Service, Motors, and Transformer Load Calculations), and other basic and advanced electrical classes at Bossier Parish Community College, Louisiana State University-Shreveport, Louisiana Technical College (Forcht Wade Correction Center), and Southern University-Shreveport to include serving as the Department Head of Industrial Electricity at Houston Community College-Stafford, Texas.

Mr. Walker is best known is for his hands-on approach and the ability to simplify and explain the most difficult electrical subject matters. He is a master electrician and holds a Louisiana state license as an electrical contractor. He has a degree in electrical engineering from the University of South Carolina and has worked as an electrical engineer for E.I. DuPont and Westinghouse at The Savannah River Plant (Company) of Aiken, South Carolina and M.W. Kellogg of Houston, Texas.

In his daily life Mr. Walker is a devoted Christian who has a passion for serving Christ, his fellowman and teaching and spreading the Word of God. As a recipient of three honorable discharges, he served over 9 years in the United States Army.

He enjoys traveling, wood-works and carpentry but is best known for his famous smoked barbeque ribs and sweet ice tea.